ULTIMATE

SECURITY

To Kalol,
with finest memories

ULTIMATE SECURITY

The Environmental Basis of Political Stability

—◆—

NORMAN MYERS

ISLAND PRESS

Washington, D.C. ■ Covelo, California

This edition published by arrangement with W.W. Norton & Co., Inc.

Library of Congress Cataloging-in-Publication Data

Myers, Norman.
  Ultimate security: the environmental basis of political stability/ Norman Myers.
      p.   cm.
  Originally published: New York: W.W. Norton, 1994.
  Includes bibliographical references and index.
  ISBN 1-55963-499-5 (pbk.)
  1. Environmental policy—United States. 2. National security—Environmental aspects—United States. 3. Environmental policy—Case studies.   I. Title.
  GE180.M94   1996
  363.7'00973—dc20   96-33559
                                                              CIP

Printed on recycled, acid-free paper ✪

Manufactured in the United States of America

10 9 8 7 6 5 4 3 2 1

Contents

CONTENTS

CASE STUDIES: GLOBAL EXAMPLES

THE NEW SECURITY

Preface

In March 1996 I keynoted a gathering of several dozen security experts in Washington, D.C. They were assembled to discuss what has come to be known as Environmental Security. They included individuals from the Pentagon, the State Department, the White House, the CIA, the National Security Council, and Congress. Among the Pentagonites was a person who briskly argued for the new line of defense analysis labeled Environmental Security. Shouldn't military planners now calculate whether a single further F-15 fighter plane at $125 million would purchase more real, enduring, and all-round security than the same sum spent on pushing back the deserts, replanting the forests, protecting farmland soil, stabilizing climate, slowing population growth, and a lengthy list of similar items? The Pentagonite made a good case, as befitted a person running a new department with the specific designation of Environmental Security.

It is itself a coup that top military brass were actually arguing that the biggest security threats may now be non-military threats, to be countered by non-military means. I had never thought I would raise a cheer for the Pentagon view of life. Hallelujah, it seems there is no limit to what enlightened people can envision. But then I thought of the Congressional slashers of the United States' foreign aid budget in the fall of 1995, particularly their 35 percent cutback in funding for population planning. As a result, almost 7 million couples in developing countries will be denied birth-control facilities during the next fiscal year, and in turn, there will be 4 million unintended pregnancies, 1.9 million unwanted births, and 1.6 million abortions. In return the United States will save $191 million, or the equivalent of eight hours of military spending by the Pentagon.

This Congressional hiccup notwithstanding, I reflected on how far we have come in such a short time. When I first started publishing papers on environmental security ten years ago, the topic was viewed by many experts as an esoteric digression from real world problems. "Yeah, air pollution is important, but it doesn't compare with gunsmoke." Today they talk a different language as nations come within an ace of firing at each other's ships on fishing grounds of the high seas. There have been two "all but" outbreaks of violence over tuna in the northeast Atlantic, one over squid in the southwest Atlantic, and one over crab and one over salmon in the north Pacific, and one over pollock in the Sea of Okhotsk off eastern Russia. Many governments are becoming more concerned with food security than with military security, as well they might at a time when world grain stocks are at their lowest level in decades and agriculture is increasingly beset by freak weather as a likely harbinger of global warming. And yet, there are still political leaders who proclaim that they will not cede one square meter of territory to a foreign invader, while allowing hundreds of square kilometers of topsoil to wash away each year.

There are still further signs of a new understanding. United States Secretary of State Warren Christopher has learned during the Middle East negotiations that the most strategic liquid in the region is not oil but water. He has been urging his diplomat colleagues to recognize the new truths of the environmental cause. Other leaders singing the new song include Chancellor Helmut Kohl of Germany and Prime Minister Gro Brundtland of Norway. In my own country, Britain, even a few ecclesiastical leaders are starting to urge the merits of the environment—though one bishop has jibbed at the title of my book, claiming that concerns of ultimate security surely belong in the church's court.

As I re-checked through this book, I figure it was quite the most difficult of all my books to write. It is far more complex in its subject matter and makeup. Its predominant theme is environmental security, but there is a larger theme, too, little though it is touched upon in the text. It is that we all share one Earth, that we are all Earthlings, and that by virtue of the environmental unity of Earth, all our futures are all our futures together. Yet many of us still prefer to think of ourselves as Americans, Britishers, Brazilians, and Indians—even though we all breathe the same atmosphere, and we all share the same boundless and boundary-less world that is the planetary ecosystem. Willy-nilly, we are all card-carrying citizens of the One Earth. All too unconsciously, we reject the idea. As the pop group, Dire Straits, tells us, "Denial is not just a river in Egypt."

To present the book's message, I have called on a number of friends and colleagues for help and advice. They have been supportive to exceptional degree. While they are too numerous to list, I particularly want to thank Drs. Frank Barnaby, David Duthie, and Tim O'Riordan, and Ambassador Paul Boecker.

Still more helpful has been my editor at W. W. Norton, Mary Cunnane. I first broached the idea of the book to her more years ago than I care to remember. She responded with the enthusiasm she has shown for my other book ideas, and she urged me to get cracking forthwith. I took a longer look at the issues involved, whereupon I balked at the complexities of the topic overall. I then balked again, and yet again. A whole year went by as I grappled with more thematic problems than in several of my other books put together. Throughout it all Mary was a paragon

PREFACE

of patience. She encouraged and cajoled, she prompted and prodded—invariably in the friendliest way. For your support, Mary, gentle and firm at the same time, during the book's gestation, let alone it protracted writing, please accept my emphatic thanks.

Another person in New York showed similar smiling tact while the book's writing was delayed beyond many people's endurance. Ginger Barber, my literary agent, contained her soul in patience as I sent one assurance after another that things were moving along all right, the manuscript would be in the mail, all would be well. I hope I express gratitude that measures up, Ginger, to your forbearance.

In the upshot, it took me more than two years to get The Thing completed. The theme was simply too convoluted to be handled in short order. I finally reached the end thanks to a colleague and friend who, better than many of the experts whom I consulted, understood the tale I was trying to tell. She outlined the book with me, she researched it, she hunted down detail after detail, she dreamed up ideas that would not have occurred to me. And while it was the least of her contributions, she spent more time on a further activity than on the others: she typed the entire manuscript several times over. When I handed her a draft chapter to read and she responded "Yes, that's fine," I knew it worked. She has a very strong sense of what is right, and whenever she decided some chapter or other was right, then it was more than right by me. She worked with me on three earlier books, and this one turned out to be more "my book," with more of me in it, because the two of us have shared a common outlook on our world. For much more that you know you have contributed, thanks, Jennie Kent.

Now that the book is going into a paperback edition, I want to express appreciation to an editor at the new publisher, Island Press. Thank you, Jim Jordan, for being so enthusiastic about the idea, and for handling the contractual details with such agreeable aplomb.

ULTIMATE

SECURITY

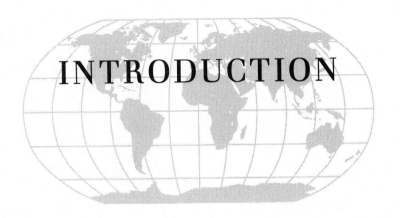

INTRODUCTION

1

What It's All About

There is a new and different threat to our national security emerging—the destruction of our environments. I believe that one of our key national security objectives must be to reverse the accelerating pace of environmental destruction around the globe.

> —Senator Sam Nunn,
> Chairman of the Senate Armed Services
> Committee, U.S. Senate, 1990

On a sparkling winter's day in December 1984, I traveled to Geneva to sign on as a senior adviser to the World Commission on Environment and Development. Sponsored by the United Nations, the commission comprised world leaders who were taking three years to review the state of the planet with regard to economic growth. Around the table I found two dozen politicians, policymakers, bankers, industrialists, trade experts, financiers, and lawyers. Not a single "eco-freak" among them. These were hard-nosed people, accustomed to making their way in the political arena and the competitive marketplace, hardly the sort of people to engage in "way-out" ideas.

When my turn came to address them, I did not dwell on my favorite themes of tropical forests, mass extinction of species, and other established topics of environmental crisis. I urged them to consider something new: environmental security. I even proposed it should become a leading item on their agenda. At first they were skeptical. But by the time I attended a second meeting, they were ready to take a second look. After another meeting, they accepted my proposition. When they got ready to prepare their report, they asked me to draft a whole chapter on the subject.

Best of all, they eventually decided to open one of their final reports[1] with one of the most far-reaching statements I have encountered from a group as sober-sided as the commission: "Humankind faces two great threats. The first is that of a nuclear exchange. Let us hope it remains no more than a diminishing prospect for the future. The second is that of environmental ruin worldwide—and far from being a prospect for the future, it is a fact right now."

Well might the commissioners have produced such a startling assertion. We face environmental decline on every side. In just the past year we have lost 26 billion tons of topsoil, enough to grow 9 million tons of food and to make up the diets of more than 200 million starving people. We have lost almost 60,000 square miles, equivalent in area to New England, of tropical forest, which has cost us dearly in terms of timber harvests, watershed services, species habitats, and climate stability. Another 23,000 square miles, the same size as Ireland, have been desertified to the extent that they won't be able to grow food again for decades at best. Tens of thousands of our fellow species have been denied living space on our One Earth. Some of them could well have supplied new foods such as the kiwi fruit (now enjoying a worldwide market worth $22 million a year) or new drugs against cancer and AIDS. During the past year, too, the ozone layer has lost more of its power to protect us from cancer-causing ultraviolet radiation. Perhaps worst of all, our greenhouse gas emissions have taken us a solid step closer to global warming. And so on and so drearily forth. The reader will be familiar with a lengthy list of environmental woes. The past year has also seen another 93 million people—equivalent to a "new Mexico"—join the global community.

Some community. It is a collection of almost 200 nations that during that same year have spent around $1 trillion on fighting one another or getting ready to do so. In dozens of wars and other outbreaks of violence, at least one-third of a million people have been killed, directly or indirectly. Many of these conflicts have an environmental source. As too many people make too many demands on too few farmlands, water stocks, and other necessities of daily life, they increasingly resort to force to ensure their share. Examples abound. In a world of growing shortages, there will be no shortage of further examples of environmentally inspired violence—whether high profile or low key, local or widespread, distant or next door, recognized as environmentally derived or not.

Many environmental problems, aside from the "biggies" such as global warming and depletion of the ozone layer, are largely located in the developing world, but they are problematic enough for many of their repercussions to spread to the ends of the Earth. They can be relieved, it was estimated at the June 1992 Earth Summit conference in Rio de Janeiro, for an outlay of $625 billion a year, $500 billion to come from developing countries and $125 billion from developed countries. Per citizen that works out to about $100, or $\frac{1}{7}$th and $\frac{1}{180}$th of annual incomes, respectively. We—especially we in the rich world who sometimes sound as if we have never been poorer—have decided thus far that it is too expensive. The total bill of

$625 billion is hardly more than double the annual defense budget of the most militaristic nation on Earth today, the United States.

A Personal Odyssey

I was first struck by thoughts of environmental security way back in the late 1970s while I was living in Kenya. I had visited the neighboring country to the north, Ethiopia, a few years before the outbreak of its 1977–78 war with Somalia, and I pondered some environmental makings of the conflict. Ethiopia's main farming area in its highlands had been regularly losing well over 1 billion tons of topsoil a year (for comparison, note that the United States has been losing 2.8 billion tons per year, from a cropland expanse twenty times as large). Farmland fertility declined, and the highlands could no longer support the peasant communities that had been growing bigger through population increase. So large numbers of Ethiopian peasants began to migrate out of the highlands, some of them toward an area known as the Ogaden on the border with Somalia—an area that had long been a source of ethnic friction between the two countries.

Mistakenly, Somalia saw the throngs of migratory Ethiopians as a threat to its frontier zone. War broke out. The Soviet Union had already come in on the side of Ethiopia, so the United States did the same with Somalia. While the superpowers played no part in the fighting, they supplied arms aplenty to their clients. The war and its aftermath eventually cost $2 billion. Subsequently it turned out that Ethiopia's environmental problems could have been cured for only one-quarter as much expense, notably through reforestation of its highlands in order to counter soil erosion. The outcome prompted the comment by an American environmental expert and former United Nations leader, Peter Thacher, that we shall frequently face a prospect of "Trees now or tanks later."

Shortly afterward, I had noted that in a lengthy list of countries, both within Africa and elsewhere, decline of the environmental underpinnings of agriculture led to rising prices for food, and sometimes to outright shortages of food. In turn, these shortages helped trigger civil disorders, insurgencies, and military eruptions. True, a number of other factors were often mixed in, notably unjust division of farmlands and poor agricultural policies, and above all the sheer pressures of population growth. But if environmental factors were less than exclusively causative factors, they were often strong contributing factors. They helped foster food riots, occasionally leading to the overthrow of governments, in countries as disparate as Egypt, Zambia, Iran, Bangladesh, Colombia, Bolivia, and Haiti, among a dozen others.

At the same time, I was aware that problems over food supplies were showing up among developed nations as well—for example, with respect to marine fisheries. In the North Atlantic, Britain and Iceland had come to the edge of hostilities over cod stocks. At least sixteen other major clashes over fisheries had occurred in other parts of the world, including those between the United States and Peru, and the Soviet Union and Japan. These disputes served as a portent, I surmised, of what could become a frequent phenomenon in the future, especially in light of the failure of fisheries around the world to produce ever-greater harvests. Fishes and friction?

Also during the mid-1970s, I met General Moshe Dayan, who played a leading part in his country's victory over the Arabs in the 1967 war. He told me that Israel unleashed its planes and tanks as soon as it learned that Syria and Jordan were moving to cut off water flows from rivers that originated outside Israel's borders. Far from being a political war prompted by the antics of Egypt's president Gamal Abdel Nasser, it was partially a water war.

In the early 1980s I visited the Philippines. I remember

standing on a mountaintop and looking out across a horizon-to-horizon expanse of slash-and-burners who were eliminating the country's remaining forests. While talking with the peasants doing this work, I was struck by the fact that they were undercutting their prospects of future livelihood through deforestation-derived erosion of topsoil from their croplands. But it was precisely their poverty that, they insisted, left them no alternative. I noticed, too, that the upland deforestation was levying fearful costs on downstream areas, smothering irrigation networks and hydropower facilities with silt and other washed-off debris, thus spreading poverty to still other communities. I also mulled over the conviction of many peasants that the government was indifferent to their plight, and that their best hope for redress—so they asserted—lay with support for rural guerrillas who already controlled entire sectors of the nation.

In addition, I pondered the thought that runaway population growth was reinforcing problems of environmental rundown in many developing countries. There had already been population-prompted violence between India and Bangladesh and between El Salvador and Honduras, and smaller skirmishes between Nigeria and its neighbors and between Rwanda and Tanzania. I had noted that in a string of African countries—Sudan, Chad, Angola, and Mozambique—there seemed to be a connection between population pressures, environmental ruin, and constant civil war.

At the same time, I had started to look into a future where global warming would bring drought and famine, plus coastal flooding, to a host of countries. Allied with overpopulation, this could trigger an outbreak of well over 100 million environmental refugees.

The more I thought about these separate instances, the more I wondered if they did not add up to some sort of big picture of security headed into basically new directions. So I wrote up my analysis in a series of articles in scientific journals.[2] Some readers reckoned they made sense; others thought they were nonsense. But they stimulated other analysts to take a poke at the question,[3] and soon the idea began to take on a legitimacy of its own. I was asked to discuss it with security experts in the U.S. State Department, the Pentagon, and other agencies of the American government. I illustrated my thesis with the observation that in Central America—a region of particular concern to the United States—it could hardly be coincidence that the country with the most devastated environments, El Salvador, was also the country with the most political instability and the most violence.

I laid out my ideas at a workshop arranged by the North Atlantic Treaty Organization, followed by discussions with defense experts in Sweden and Australia. At a conference in Russia organized by Pugwash—a worldwide network of security-focused scientists—I found that a number of Soviet military leaders were interested too. They were probably prompted by Mikhail Gorbachev's assertion that "The relationship between man and the environment has become menacing. Problems of ecological security affect all, the rich and the poor. The threat from the sky is no longer missiles but global warming."[4]

Today I find that the new concept is becoming accepted on the international scene. It has been increasingly talked about by such leaders as Prime Minister Brian Mulroney of Canada, Chancellor Helmut Kohl of Germany, President François Mitterand of France, Prime Minister Gro Brundtland of Norway, former prime minister Margaret Thatcher of Britain, the late prime minister Rajiv Gandhi of India, and former Secretary-General of the United Nations Javier Perez DeCuellar. The list also includes such foreign-policy and security experts as Robert McNamara, George Kennan, James Baker, Prince Sadruddin Aga Khan, and Sir Sonny Ramphal. All appear to go along with the thought that whereas the past forty years have been dominated by the Cold War, the next forty years will surely be dominated by environmental conflicts. They will add up to a variant of World War III, a war we are waging against the Earth—and a war we are winning hands down.

Moreover, it is a war of us all against us all. Because America's grainlands help to feed over a hundred countries around the world, soil erosion in Indiana should be of as much concern to everyone as soil erosion in India. In an interdependent world where developed nations and developing nations are deeply involved in each other's affairs (till death do them part), desertification in California and Australia counts as much as desertification in Mexico and Pakistan. Chemicals that damage the ozone layer are mainly manufactured and utilized in the Northern Hemisphere, but the nations likely to suffer in the first instance are those nearest Antarctica. Deforestation in Borneo will impinge, if only through its climate linkages, on farmers in America's great grain belt. Global warming and mass extinction of species will affect everybody everywhere. We are all in the same environmental boat, and we shall all get wet as it springs leak after leak. Not even the most advanced nation can insulate itself from environmental impacts, no matter how strong it may be economically or how advanced technologically or how powerful militarily.

Moreover, any New World Order must reflect the new situation where the principal threat to security and peace stems from environmental breakdown, plus the need for access to natural resources that are increasingly scarce as more people make greater demands upon them. If the oil wars have begun, the water wars are on the horizon, to be followed by resource wars over key environmental supports for economies right around the world. The security strategy is obsolete that focuses on military prowess as the predominant mode to safeguard individual nations' interests. Equally to the point, a New World Order must focus on the new world that embraces all nations as components of a unity far greater than the sum of its parts. Fortunately, the end of the Cold War offers us glorious scope to embark on entirely new approaches to security. True, the Gulf War has revealed a fine instance of collective response to a single nation's aggression. But the war's conduct and its aftermath, with emphasis confined to conventional notions of security, have given scant regard to the far greater challenges of the new security.

This also means that a "siege response" to international problems of the environmental sort would be decades out of date. A basic fact of the late twentieth century is that the world is finally becoming one world. For a multitude of economic as well as environmental reasons, interdependence is here to stay. It is as distant from the "hoist up the drawbridge" world of just a few decades ago as was that world from the world of several centuries ago when nations were becoming established. Yet however much interdependence is a built-in fact of our new world, we have yet to mobilize the political collaboration to reflect it: we simply do not recognize it as a strict fact of life. Central to this fact is the new understanding of security. This is not a choice we can confront at some stage in the future. The option of "business as usual" security was foreclosed a good while back on the grounds that global warming alone will cause upheavals—in the agriculture, water, forestry, fisheries, and industry sectors, among a host of other sectors—for all nations regardless of location, economic strength, or military prowess. So the only choice ahead of us is how to exploit our new situation to best advantage all around.

The changed outlook will not come easily. As a leading scientific commentator remarked recently, the two most important features of our new world have nothing to do with conventional politics or economics, least of all with military strategy. They are, first, that no territory can support an indefinite increase either in its number of creatures or in consumption per creature, let alone both; second, that all mainstream policies of all governments assume that, on the contrary, it can. Together these facts add up to the great global threat, and political strategies and security planning alike will have to adapt radically to meet it.

The Realpolitik Skeptic

To reiterate a central point, the new concept evokes doubt—
and then some—among certain security experts. I remember a
conversation in 1990 with a senior member of Britain's foreign-policy establishment. We were talking about sub-Saharan
Africa and, hard-nosed analyst that he was, he proclaimed that
if the region declined into a vortex of poverty, instability, and
despair, what was that to Britain? How would it make a difference to the nation's true interests? There was no longer the
argument, he asserted, that the region contained abundant
stocks of critical minerals; they could be stockpiled, or they
could be substituted through technology. Trade opportunities
were no big deal, he went on; Britain's main trading partners
were the other developed nations, and in any case the region's
trade had declined to only one-quarter of what it was in 1960.
Nor did the region influence strategic shipping lanes, as used to
be the case when the Cape of Good Hope was seen as vital to
the West's trade routes. Nor was he concerned that an impoverished and unstable Africa would offer splendid scope for military adventurism on the part of Khaddafi-style mavericks. Environmental refugees: how could they make their way to
Britain's shores?

To all my arguments, the official affected a realpolitik indifference. He rejected notions of any British interdependence
with sub-Saharan Africa—and, for that matter, most kinds of
global interdependence as conventionally perceived, whether
interdependence of one nation with other nations today or interdependence of one nation today with all nations of the future. He appeared to overlook emergent forms of environmental interdependence that could affect Britain's interests in
profound fashion over the long term. What if Africa's forests
continue to be torched, releasing huge amounts of carbon diox-

ide and other greenhouse gases? What if the demise of those forests affects global climate in other ways that we do not know much about yet, such as shifts in the planet's albedo balance (albedo being the "shininess" of the Earth's surface, a prime determinant of climate)? What if sub-Saharan Africa steadily becomes Saharan Africa, with all the climatic dislocations that could stem from broad-scale desertification and desiccation of an entire segment of Earth's atmosphere?

Finally I offered the case that the world could not feel secure in a proper broad sense as long as a major part of the global community was falling further and further behind, becoming the Third World of the Third World. On humanitarian grounds alone, should we not hold out a helping hand? After all, this response seemed to carry weight in 1985 when Bob Geldof and his Liveaid concert attracted one-quarter of all rich-world people to tune in to his message via their television sets, and then to reach into their pockets. In Britain, the money raised through this one appeal to personal charity was more than the public funds the government had been planning to spend on Ethiopia. What clearer statement could there have been that people cared simply because they cared? In the long shot, public policy would have to be grounded in public opinion if it were to remain a legitimate expression of the nation.

To make the point more specific and, I hoped, more appealing to the foreign-relations expert, I raised the question of "the children's holocaust." Each year 13 million children die in the developing world, half of them in Africa, from easily preventable diseases. To put the figure in perspective, we can reckon it is the equivalent of a jumbo jet full of children hitting the ground every quarter of an hour, day and night throughout the year. When parents see such vast numbers of their children dying they are inclined to produce extra-large families as a form of "insurance." During the 1990s we could save 100 million developing-world children for a cost of only $500 million. Even if the rich nations were to foot the whole bill, it would work out to a mere half dollar per citizen per year. No other human generation has ever had a chance to save so many children so cheaply. Should we not leap at the chance?

Even in the face of this argument, my diplomat friend was unmoved. He spoke of sundry blockages to the idea, such as "compassion fatigue" on the part of developed-world voters. What political muscle is there to moral imperatives?

And yet, I thought, and yet. While African children are dependent upon us developed-world citizens for at least part of their survival prospects, we are dependent upon them, too, for our sense of "belonging" as full-blown members of global society. However little we may recognize it, humankind has grown to the stage of being a single community, irrevocably interlocked in each other's affairs and needs. We live with each other, for each other, through each other. We can no longer live apart from each other; we are a part of each other. Does this not offer splendid opportunity for us to show we now deserve to rank as "human kind"? Or to pitch the question of community on a new level, isn't there a sense of individual wellbeing that reflects our relationships with each other—with all the "each others" right around the world? Isn't there a fresh type of security that also counts, an inner security that ultimately forms the bedrock of our being?

This, I feel, is the clincher argument. In comparison, all the others pale into also-ran status, as well they should. This book will present many cold-eyed analyses of trade-offs, clinical economics, and political machinations. But in the end there is only one basic point, that nobody can feel finally secure as long as others are persistently insecure. The reader will find the point rarely raised again in this book. What's the purpose? Either you accept it or you don't. Many experts, whose opinions I respect, will object that this is not the way the world works. I believe it should be and could be.

2

Environmental Security: How It Works

The environmental problems of the poor will affect the rich as well, in the not too distant future, transmitted through political instability and turmoil.

—Gro Harlem Brundtland,
Prime Minister of Norway, 1986

Environmental problems can figure as causes of conflict. If we continue on our road to environmental ruin worldwide, they will likely become predominant causes of conflict in the decades ahead. That is the essential message of this book. In the chapters that follow, we shall look at case studies to illustrate the theme that environment has become a fundamental factor in security issues in many regions already, and in the future will increasingly lie at the heart of security concerns of nations around the world.[1]

Example: water in the Middle East. This vital liquid is the cause of much pushing and shoving among countries of the region. In fact, there have been conflicts over Middle East water for thousands of years, and they will likely continue long after the oil wells run dry. For instance, Israel and Jordan depend on the Jordan River for much of their water. The two countries' need for the river's water in the year 2000 is expected to be 130 percent of the total flow. Somebody is going to run badly short of what is ultimately the most precious of all resources in the Middle East. King Hussein of Jordan has already asserted that water shortages are the only reason his country should go to war again with Israel.

Similar water disputes have arisen between India and Pakistan because of diversion of water from the Indus River, and between India and Bangladesh over the Ganges River for much the same reason. Further water-supply problems have erupted in the river basins of the Mekong, Nile, and Rio de la Plata. Of more than 200 major river systems, around 150 are shared by two nations, and more than 50 by three to ten nations. All in all, they supply nearly 40 percent of the world's population with water for domestic use, agriculture, hydropower, and similar purposes. Between 1940 and 1980, water consumption around the world has more than doubled, and it can be expected to double again within another twenty years. Note, too, that two-thirds of the water will be used to produce that basic commodity of people everywhere, food. In southern Asia, water shortages look likely to curtail plans to expand agriculture, and as many as 80 countries, with two-fifths of the world's population, already suffer serious water deficits. So water, scarce and precious water, could soon become a cause célèbre of conflict between nations as fast-growing numbers of people make ever-increasing demands on this crucial resource. Remember, they ain't making any more of the stuff.

We can discern other linkages between environment and conflict with respect to deforestation. In the Ganges River system, dependent as it is on tree cover in the catchment foothills of the Himalayas, monsoonal flooding has become so widespread that it regularly imposes damages to crops, livestock, and property worth $1 billion a year among downstream communities of India and Bangladesh, even though the main deforestation occurs in another country altogether, Nepal. The result is deteriorating relations among the three governments; India shouts at rather than speaks to Bangladesh, and it has recently broken off diplomatic ties with Nepal.

As yet another example of linkages between environmental status and global conflicts, desertification generates broad-scale problems for human welfare and political stability. In the Sahel zone of Africa, not a single government survived the droughts of the 1970s and 1980s, several fell twice over, and a few, like that of the Sudan, are moving toward a third collapse.

In the instances listed, the linkages are readily apparent. In other cases, the impact is more deferred and diffuse, as in the case of species extinctions and gene depletion, with all that will mean for genetic contributions to agriculture, medicine, and industry. Probably the most deferred and diffuse impact of all, although the most consequential all around, will prove to be that of climatic dislocation. Buildup of carbon dioxide and other greenhouse gases in the global atmosphere will, if continued as projected, engender far-reaching disruptions for temperature and rainfall patterns. As a result of probably warmer and drier weather in its established grain belt, leading to severe drought persisting year in and year out, the United States could well become less capable of growing food. Conversely, Russia and Ukraine, possibly enjoying more favorable circumstances in a good part of their territories, could become major suppliers of surplus food. India could conceivably find itself better off in agricultural terms and Pakistan worse off, which in turn could affect the relations between these two traditional adversaries. There will be many other "winners" and "losers" in a greenhouse-affected world, with all manner of destabilizing repercussions for a world already experiencing other types of environmental turmoil—hence we shall have a world in which we all end up losers.

The New Security

In short, there is an array of environmental factors that contribute to problems of security—security in its proper all-round sense, security for all, security forever. (For an overview of what I mean by security itself, see the box at the end of this chapter.) Each environmental factor serves, to one degree or another, as a source of economic disruption, social tension, and political antagonism. For sure, certain of the linkages are diffuse in their workings and hence difficult to discern in their immediate operation. But they are real and important already, and they are fast growing in their number and extent. While they may not always trigger outright confrontation, they help to destabilize societies in an already unstable world—a world in which we can expect this destabilizing process to become more common as growing numbers of people seek to sustain themselves from declining environments.

It all adds up to a new fact of global living. Security concerns can no longer be confined to traditional ideas of soldiers and tanks, bombs and missiles. Increasingly they include the environmental resources that underpin our material welfare. These resources include soil, water, forests, and climate, all prime components of a nation's environmental foundations. If these foundations are depleted, the nation's economy will eventually decline, its social fabric will deteriorate, and its political structure will become destabilized. The outcome is all too likely to be conflict, whether in the form of disorder and insurrection within a nation or tensions and hostilities with other nations.

This is not to deny that natural resources have often been important for the security of individual nations in the past, even to the extent of generating conflict.[2] Recall Hitler's supposed urge for sheer living space. In more recent times there have been the Algerian war of independence, the Congo civil war, the western Sahara revolt, and most recently, the Gulf war. This last conflict was primarily about the strategic resource of oil rather than about the independence of Kuwait. In each of these conflicts, access to natural resources was a salient if not the primary consideration. But the resources were usually minerals (especially fossil fuels), thus the conflicts contrast with more recent disputes over water, agricultural lands, and the like. There is a still bigger difference today, moreover, in that environmental factors look likely to become pervasive and even predominant as sources of conflict. So there is a quantum advance in the scale of environment's role in conflict. Nations everywhere will likely suffer if much of the world is impoverished through environmental ills of myriad sorts, and they will certainly suffer as the entire world is overtaken by global warming.

All in all, then, national security is no longer about fighting forces and weaponry alone. It relates increasingly to watersheds, croplands, forests, genetic resources, climate, and other factors rarely considered by military experts and political leaders, but that taken together deserve to be viewed as equally crucial to a nation's security as military prowess.[3] The situation is epitomized by the leader who proclaims he will not permit one square meter of national territory to be ceded to a foreign invader, while allowing hundreds of square miles of topsoil to be eroded away each year.

What Also Counts

Of course we must be careful not to overstate the case. Not all environmental problems lead to conflict, and not all conflicts stem from environmental problems. Far from it. But there is enough evidence, as we shall see in this book, for the central thesis to stand. Similarly, while environmental phenomena contribute to conflicts, they can rarely be described as exclusive causes. There are too many other variables mixed in, such as inefficient economies, unjust social systems, and repressive governments, any of which can predispose a nation to instability (and thus, in turn, make it especially susceptible to environmental problems). In developing countries, poverty—absolute poverty, meaning a lack of the essentials of acceptable life, such as food, water, shelter, and health—afflicts well over one billion people, or one-fifth of humankind. Their numbers grow larger with every passing year. Impoverished people become desperate people, all too ready to challenge governments through, for example, support of guerrilla groups as in the Philippines and Peru. At the same time, impoverished people feel driven by their plight to overwork their croplands, to clear forests, and to cultivate arid lands and mountain slopes for additional croplands, all of which trigger soil erosion and other environmental ills, and result in poverty compounded.

In developing countries there often are a number of further factors that undermine security. They include faulty economic policies, inflexible political structures, oligarchical regimes, oppressive governments, and other adverse factors that have nothing directly to do with environment. But as we shall see in the case studies that follow, these deficiencies often aggravate environmental problems, and are aggravated by environmental problems in turn.

The biggest factor of all in many developing countries is the population explosion, still to enter its most explosive phase. There are now 4.3 billion people in developing countries, half as many again as just thirty-five years ago. The number is projected to surge to 7.2 billion, 67 percent more than today, by the year 2025. As we shall see, population pressures already generate discord and strife of multiple forms, often erupting in violence. Equally to the point, these pressures encourage the overexploitation of environments such as farmlands and water stocks, a process that fosters the spread of poverty among the "bottom billion" people.

These poorest of the poor cause a disproportionate share of environmental degradation. They find they have no alternative. Their concern is not tomorrow's world, it is tonight's supper. Constituting 20 percent of the world's population, they enjoy about 1 percent of global income, trade, investment, and bank lending. Obliged to live off nature's capital, gross national product is less relevant than gross natural product, and shortage of biomass is worse than shortage of cash.

While these bottom billion often cause more environmental decline than the other three billion developing-world people put together, the world's top billion also cause an undue amount of environmental decline. With less than one-quarter of the world's population, and through their excessively consumerist and wasteful life-styles, they are responsible for most of the overuse of raw materials: they account for 70 to 85 percent of the world's consumption of fossil fuel and the manufacture of chemicals, as well as military spending. Most important of all in the long run, they cause the great bulk of ozone-layer depletion and global warming, two leading problems that will grievously harm the prospects of people throughout the world.

In short, there is a growing connection between environment and conflict. Environmental deficiencies engender conditions which render conflict all the more likely. These deficiencies can serve to determine the source of conflict, they can act as multipliers that aggravate core causes of conflict, and they can help to shape the nature of conflict. Moreover, not only can they contribute to conflict, they can stimulate the growing use of force to repress disaffection among those who suffer the consequences of environmental decline.

Collective Security

A further crucial lesson emerges from the new view of environment-based security. While we need to expand our understanding of security to incorporate an environmental dimension, we also need to adapt our policy responses by placing greater emphasis on collective security. The challenge is well illustrated by the question of carbon dioxide and other greenhouse gases. This is a problem to which all nations contribute, by which all will be affected, from which no nation can remotely hope to insulate itself, and against which no nation can deploy worthwhile measures on its own.

What can nations do to meet the new challenges? Primarily they can recognize that many forms of environmental impoverishment constitute a distinctive category of international problems, unlike any of the past. These new problems lie beyond the scope of established diplomacy and international relations. While impinging on the strategic interests of individual nations, they prove altogether immune to the standard response to major threats, namely, military force. We cannot launch fighter planes into the sky to resist global warming, we cannot dispatch tanks to counter the advancing desert, we cannot fire the smartest missiles against the rising sea.

These problems require a response different in yet another sense. This response must emphasize cooperation rather than confrontation within the international arena. No nation can meet the challenges of global change on its own. Nor can any nation protect itself from the actions—or inaction—of others. Plainly this demands a seismic shift in spirit and strategy alike. It postulates as big a change for the nation-state as any since the emergence of the nation-state four hundred years ago. To cite Sir Crispin Tickell, former British ambassador to the United Nations, "No man is an island, no island is an island, no continent is an island. Yet nation-states still think principally if not almost entirely in terms of islands—economic, political, environmental islands." And to cite the opening sentence of the Brundtland Commission's report, "Our Earth is one, our world is not." Hitherto we have adopted a stance that essentially says that what I gain must be what you lose, and vice versa. Today, for the first time and for all time henceforth, we face situations where we shall all win together or we shall all lose together.

Developed Nations and Developing Nations

Certain of the instances cited—notably water shortages, agricultural decline, and deforestation—are located mainly in developing nations. Why, then, the reader might ask, should security analysts in developed nations be concerned with water disputes in the Indian subcontinent, food shortages in Africa, and deforestation in Amazonia? There are several reasons.

First is that the developed world has a decisive stake in the wellbeing of the developing world—little though that connection may be recognized in certain councils of power. In particular, the United States finds that its prime hope for expanding its exports lies with developing countries. By the mid-1980s a full two-fifths of exports were going to those countries (and they were accounting for one in three of American manufacturing jobs). The proportion of exports to developing countries could swell to one-half as soon as the year 2000,[4] which would mean that half of U.S. foreign-exchange earnings and half of the balance-of-trade successes (or problems) would be riding on markets in Latin America, Asia, and Africa. But this latter prospect depends upon the capacity of developing countries' economies to keep on growing. In turn, this depends in major measure upon the environmental-resource base in which the economies of many developing countries are grounded. If that base becomes depleted, the economies will falter and fail, becoming less able to afford America's exports. In the single year of 1985, economic stagnation in developing countries caused a drop of $3 billion in American exports and the loss of 220,000 jobs in exports-oriented industries; the decade-long recession in the developing world cost 1.4 million jobs.[5]

To this extent, the economic health of the United States is tied to the environmental health of developing countries. Developing-world poverty has become a luxury we can no longer afford. Clearly the economic wellbeing of the United States is a leading item in the nation's security interests. Significantly, then, though less clearly at first glance, the security of the United States reflects the environmental state of developing countries. This linkage is particularly important for those many developing countries—we shall look at several examples in this book—that represent salient economic and security interests of the United States. (Much the same applies to developed nations generally.) To cite former secretary of state George Shultz, "There can be no enduring economic prosperity for the United States without sustained economic growth in the Third World. Security and peace for Americans are contingent upon stability and peace in the developing world."

A second security linkage arises with respect to political stability in developing countries. To cite a 1988 report of the Ikle and Wohlstetter Commission on Integrated Long-Term Strategy,[6] "Violence in the Third World threatens our interests in a variety of ways. It can imperil a fledgling democracy (as in El Salvador), increase pressures for large-scale migration to the United States (as in Central America), jeopardize important U.S. bases (as in the Philippines), and threaten vital sea lanes (as in the Persian Gulf)." Again, these security linkages between the United States and developing nations apply to other developed nations, albeit with differences in accord with particular strategic relationships.

A third linkage between the developed world and the developing world lies with the fact that environmental problems in one country often spill way beyond its borders. As Brazilian Amazonia goes up in flames, we might recall that a full 30 percent of the chief greenhouse gas, carbon dioxide, already comes from the burning of tropical forests—and the proportion is rising rapidly. Another important greenhouse gas, methane, is generated primarily by the cattle and rice paddies of developing-world Asia. Thus everyone in the developed world has an emphatic interest, whether they are aware of it or not, in what goes on in developing countries around the back of the Earth. Climatic patterns embrace the whole world, and winds carry no passports.

Fourth, we are finding that environmental degradation in developing countries, accentuated by population growth and poverty, sometimes triggers mass migrations of people who can no longer sustain themselves in their erstwhile homelands. Already these desperate throngs total well over 10 million people, as many as all political and other conventional refugees combined. This could be small potatoes, however, as compared with the multitudes that could feel driven to flee their environmentally ravaged homelands in the future, notably as a result of global warming. As we shall see later in this book, they could eventually number well over 100 million. Indeed, whole waves of destitute humanity washing around the world could soon start to pose entirely new threats to international stability. This would be especially the case if the refugees feel they can best find sanctuary in developed nations. They would do this partly because developed nations offer the most prospect of support, and partly because these nations could rightly be seen as the principal cause of the main source of gross environmental decline, global warming.

Our Global Experiment

We do not know that all these unfortunate environmental out-comes will come to pass. Nor do we know that they will not come to pass. What we do know is that we are now intervening in the most basic of our planet's workings, and doing it with increasing ingenuity and vigor. At the time of the Stockholm Conference on the Human Environment in 1972, the scientific experts and the political leaders of the world came up with a lengthy list of environmental problems that merited urgent action. If we did not get on top of these problems before they got on top of us, there would be a hefty price to pay. Yet missing from that list were such altogether unrecognized items as acid rain, ozone-layer depletion, and global warming. What new items are we overlooking today? What fresh threats are roiling away under the surface, working their disruptive way, building up momentum until, by the time they make them-selves all too plain, we shall find it difficult to come to grips with them, if indeed we can come to grips with them at all except at exceptional cost and after much irreversible injury?

Were environmental problems to strike us like a heart at-tack, we would rush our ecosystems into intensive-care units and have them restored. Instead they are like a cancer, quietly undercutting our foundations, unseen and unresisted, until they eventually burst forth with deep damage all too apparent. Or, to shift the metaphor, they are like the frog in the experi-ment. If you drop a frog into a saucepan of boiling water, it will respond to its acutely and suddenly hostile environment by hopping out. If you then put the same frog into a saucepan of cold water and set the pan on the stove, the frog will enjoy its benign environment for a while. It will even think its environ-ment becomes still more benign as the water warms up. It will swim around placidly, oblivious to the threat of the rising temperature. As the water gets hot, it will start to feel drowsy. Finally it will succumb to the heat and go into a coma before boiling to death. For us, it is the world outside the window that is steadily heating up. Will we be in time to recognize the signs?

No doubt about it, we are conducting a global-scale experiment with Planet Earth. Not only will our experiment have profound repercussions, but these repercussions will extend far into the future. If nothing else will give us pause, the time scale should. Suppose that we were to continue on our environmentally disruptive way till the year 2000, then were to halt completely. Our poisoning of the planet with pollutants could not be reversed, through a mixture of natural processes and human management, within less than several decades. Desertification and ozone-layer depletion could not be fixed up in less than one century; soil erosion and tropical deforestation, several centuries; global warming, a whole millennium; and mass extinction of species, at least one million years.

Such is our experiment. It is entirely unplanned, and we know next to nothing about its outcome except that probably it will be irreversible and certainly it will be adverse. Yet we pursue our experiment with ever greater commitment, inducing we know not what results. In a situation of such overwhelming ignorance, the prudent course is to play safety first. As in other circumstances where the central factor is uncertainty, it will be better for us to find we have been roughly right than precisely wrong.

Our New Departure

So, we stand on the verge of an entirely new set of security issues. This new direction amounts to a watershed in the course of human affairs, ever since people first stared eye to hostile eye or gazed at each other with a spirit of "let's make common cause." It presents unprecedented threats to the human enterprise, and it offers unprecedented promise for us to finally build that citadel on the hilltop. If we disregard it, we shall find we have imposed a top-dollar price on ourselves. We shall have engaged in a policy of scorched earth for us all, laid waste to the Earth beyond recognition, left the Earth devastated for millennia. If, by contrast, we seize the positive option it also supplies, we shall find ourselves engaged in a joint endeavor that will, for the first time, make our world one. Viewed in its proper complete perspective, we shall confront a creative challenge without equal since humans became humans.

This book aims to set out the nature and scope of environmental security. I will show that what is most critically needed as we head into unknown territory is the vision to sense the fundamentally new dimensions of one-world living. If we accept the revolution that is under way, a revolution that can bring peace beyond dreams, we shall avoid the many violent revolutions that will surely erupt through default. Fortunately, we enjoy a "thinking dividend" with the end of the Cold War, heralding an era when old patterns are willy-nilly giving way to fresh approaches, as witness the collective response to the Gulf War. Can we expand our vision to take in the far larger problems *and* opportunities supplied by the far larger demands of environmental security?

The bottom line is that we face a time of breakdown or breakthrough. We can allow our global environment to be devastated until it scarcely functions any longer as a habitat for humankind. Or we can accept that we can make peace with one another only by making peace with the Earth.

WHAT IS THIS
THING CALLED SECURITY?

Just as health is more than the absence of disease, so security—let alone peace—is more than the absence of hostilities. Yet while a nation knows what insecurity is, it cannot say so readily what it means by security—just as disease is more readily recognizable than health in the full and proper sense of the term. We need a clear-cut idea of what security is: what it amounts to, where we get it, and, most of all, what it means to *feel secure*, whether on the part of governments, global society, communities, or individuals. At the same time, we need to determine how we can gain security without resorting to force.

In essence, and little though this is generally recognized by governments, security applies most at the level of the individual citizen. It amounts to human wellbeing: not only protection from harm and injury but access to water, food, shelter, health, employment, and other basic requisites that are the due of every person on Earth. It is the collectivity of these citizen needs—overall safety and quality of life—that should figure prominently in the nation's view of security.

As conventionally understood, a nation's view of security entails ensuring its territorial integrity and maintaining its position in the world beyond its borders. So a nation must enjoy assured access to raw materials and energy, to trading opportunities wherever available, and to scope for its government, large enterprises, and other major institutions to pursue their activities without let or hindrance. In addition, a nation also generally needs to foster its political ideals and its cultural outlook, both within its own land and further afield. Without these essentials, a nation has no security.

So, too, the entire community of nations, indeed all humankind, needs to enjoy security in the form of acceptably clean (unpolluted) environments, supplies of environmental goods such as water and food, and a stable atmosphere and climate. In short, all nations need a planetary habitat that is secure in every down-to-Earth respect—which means, in turn, that "we" are only as safe as "they" are.

However obvious all this might seem in principle, it often receives short shrift in practice. As Lester Brown, president of the Worldwatch Institute in Washington, D.C., has put it: "The gap between what we need to do to protect our environmental support systems and what we are doing is widening. Unless we redefine security, recognizing that the principal threats to our future come less from the relationship of nation to nation and more from the deteriorating relationship between ourselves and the natural systems and resources on which we depend, then the human prospect could be a bleak one. If we do not act quickly, there is a risk that environmental deterioration and social disintegration could begin to feed on each other."[7]

To be specific, let us see how all this applies to the leading nation on Earth, the United States. In terms of territorial integrity, the United States has already lost control over its borders because of mass migration, largely illegal, from south of the Rio Grande. Similarly, one aspect or another of U.S. security—its economic welfare, its trade relationships, its democratic values, its humanitarian ideals, its very climate—can be threatened by soil erosion, acid rain, water shortages, runaway growth of both population and poverty, ozone-layer depletion, mass extinction of species, global warming, and other environmental problems, in whatever part of the Earth.

In turn, this all translates into specific policy goals for the United States vis-à-vis the world at large. These goals include maintaining peace in critical regions, maintaining an open global economy with particular respect to U.S. access to important markets and key resources, safeguarding the political and economic stability of American allies and of nonaligned nations, preventing undue expansion of nations and influences opposed to American values, and fostering orderly relations among the community of nations.

These are long-established goals of U.S. interests, readily recognized. Today we need to add another goal that complements and reinforces the others. It is protecting the global environment, thereby promoting sustainable development throughout the world (in developing and developed nations alike), in order to enhance those economic and political processes that will assure a secure environmental foundation for all. This further goal will help maintain stability in international relations at a time when environmental problems threaten the stability of the entire world.

This last goal is far from widely accepted as yet. But it deserves to be considered an explicit part of U.S. security. As the world's leading economic power, the United States has a preeminent stake in the environmental health, as in the economic welfare, of the entire community of nations. In other words, it has a solid interest in such remote-seeming activities as deforestation in Latin America, desertification in Africa, soil erosion in Asia, mass extinction of species throughout the tropics, pollution of the marine realm, and climatic change wherever it arises.

This, then, is the One Earth perspective. Within this context there is the further question of economic advancement for people in developing nations who suffer absolute poverty. As we shall see throughout this book, much of this poverty is due to environmental impoverishment, which in turn deepens traditional poverty. And poverty is a potent source of insecurity. The connection was described a quarter of a century ago by Robert McNamara when he became head of the World Bank: "There can be no question but that there is a relationship between economic backwardness and violence. Our [the United States'] security is related directly to the security of the developing world. . . . Security is development, and without development there can be no security."[8]

CASE
STUDIES:
REGIONAL
EXAMPLES

3

The Middle East

Long after oil runs out, water is likely to cause wars, cement peace, and make and break empires and alliances in the region, just as it has done for thousands of years.

—John Cooley,
Former U.S. State Department Official, 1984

During twenty years of environmental work that has taken me to eighty countries, I have been constantly struck by the fact that one resource lies at the heart of all life and most livelihoods for the planet's citizens. It is water. Water once set the stage for the evolution of life, and today it is as essential to life as it ever was. It has to rank among our most precious natural endowments. It serves our needs at dozens of points in our daily rounds. With water there is survival; without it there is no food, no sustenance of any sort. No wonder that societies from ancient times attached special horror to the crime of poisoning wells. Yet for all that water is vital, many of us take it for granted. We shall soon find ourselves changing that view. Our future will be deeply compromised unless we learn to manage water as a critical ingredient of our lives.

In the Middle East in particular, water, or rather the lack of it, has for millennia weighed upon the daily lives of people throughout the region (Figure 3.1). As far back as 6500 years ago, Lagash and Umma went to war over water. There are numerous Old Testament references to the part played by water in desert societies, and the way they viewed it as a bountiful gift from God. Shortages of this most crucial of all liquids will soon become still more acute.[1] As we have already seen, Israel started the 1967 war in part because the Arabs were planning to divert the waters of the Jordan River system.[2] Israel remains on the Golan Heights and the West Bank in major measure because it wants to safeguard its access to the river system's flows, and to the water stocks of underground aquifers. At the same time, the Jordan River and other water supplies lie at the heart of the Palestinian question insofar as a large number of Palestinians live in the West Bank and find a disproportionate part of their traditional water sources now diverted for Israeli use. When Israel invaded Lebanon in 1982, it was prompted to do so in part to secure access to the Litani River. There are conflicts in the making, too, with respect to water for Turkey, Iraq, and Egypt.[3] In hardly any part of the world are geopolitics more closely linked to a key environmental factor than is the case with water in the Middle East.

Most of the region receives only 10 or 12 inches of rainfall a year, and most of that is kicked straight back into the sky by evaporation. So water for agriculture has to come mainly from rivers via irrigation, plus some groundwater. Because there is not enough irrigation water, most countries have to import a

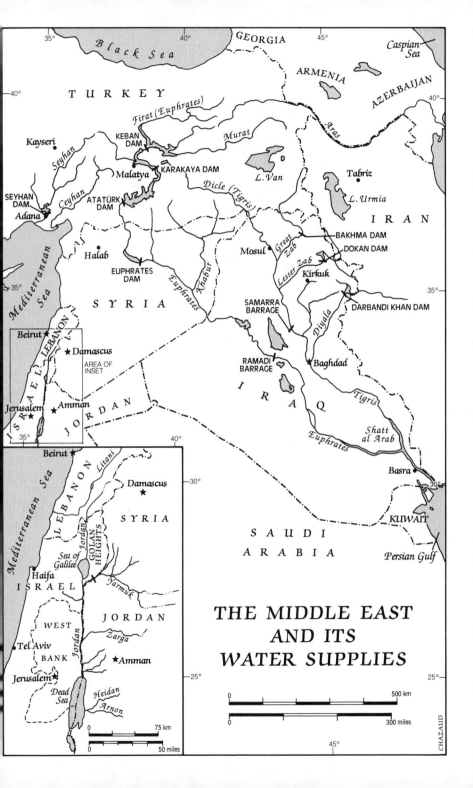

THE MIDDLE EAST
AND ITS
WATER SUPPLIES

good deal of their food; for instance, Jordan imports three-fifths of its food supplies, though it can scarcely afford the foreign exchange to do so. Moreover, most Middle East countries have population growth rates of around 3 percent per year, meaning that human numbers would double every twenty-three years or so—whereupon water demand would double too, plus a further increase to take care of rising expectations.

Of course there are other factors involved. While economic advancement usually results in greater need for water per person, it can also foster more efficient use of the resource through recycling. But even if Middle East countries deploy the most modern recycling technology, and even if they mobilize every last drop of water with maximum efficiency, population growth will still mean they will have to import ever-larger amounts of food—and the non–oil-exporting countries will be hard pressed to do so. Jordan's population is projected to increase by 73 percent by the year 2010, Syria's by 80 percent, Israel's by 30 percent, and the West Bank's by 41 percent. If the communities of the Jordan valley are already thirsty, they are going to become a great deal more so. Let us take a detailed look at the situation and its rivalries.

The Jordan River

Of all major river basins in the Middle East, the smallest is that of the Jordan River. It is also the site of the greatest tension.[4] While Israel gains about 60 percent of its water from the Jordan River and its tributaries, only 3 percent of the basin lies within the nation's pre-1967 territories. Naturally enough, Israel feels touchy about the river and who controls it, especially since the country consumes five times as much water per head as its less industrialized and less intensively farmed neighbors.

True, Israel makes good use of its water. It has pioneered the agricultural technique of trickle-drip irrigation, otherwise known as the Blue Revolution, which administers drops of water directly to the roots of crop plants. In addition, the country recycles an unusually large amount of water. But there is limited scope to do better without massive investments.[5] Israel could also exploit more of its groundwater stocks, but its proposal to increase pumping from the West Bank's underground aquifers, already supplying two-fifths of the country's water, has prompted sharp protests from West Bank Arabs and Jordan. Israel already takes nine-tenths of all water abstracted from the aquifer; nationwide it enjoys an average water consumption per person that is more than twice as much as that of the Palestinians on the West Bank.[6]

Jordan gains an even larger share of its water, three-quarters, from the river and its tributaries. It hopes to take even more, mainly to boost its agriculture. Yet even with American-level agronomic inputs, the country will be unable to feed itself from its own land by the year 2000,[7] meaning it will have to purchase large amounts of food from outside. Between 1974 and 1990 its cereal imports jumped from 171,000 to 1,491,000 tons; yet in 1990 its terms of trade were worse than in 1980. Moreover, servicing of Jordan's external debt of $7.7 billion in 1990 consumed almost one-quarter of annual export earnings, meaning there is less money available for purchasing food. Worse still, Jordan's population growth rate of 3.4 percent per year, one of the highest in the world, means the 1993 population of 3.7 million people is projected to increase two and a half times by 2025.

Hence, the country aims to grow more of its own food, almost entirely through extra irrigation. There is much potential for this approach. Jordan irrigates less than one-tenth of its croplands, a smaller proportion than any other country in the region; by contrast, Israel irrigates two-thirds of its croplands. But an expansion of irrigation will require a vastly increased supply of water. Indeed, Jordan aims to step up its water consumption by a full one-half during the brief period 1987–2002, and this drives the country's emphatic interest in its number one source of water, the Jordan River system.[8]

All in all, Jordan's water needs are projected to exceed supply by one-fifth as soon as the year 2000. By the same date, Israel could face deficits of almost one-third.[9]

The upper part of the Jordan River is already exploited for human use to its maximum capacity. True, it still has a partly tapped tributary in the Yarmuk River. But Jordan plans to build a dam, to be known as the Unity Dam, at Maqarin on its border with Syria in order to gain more water from the Yarmuk. At the same time, Syria (also with an exceptionally high population growth rate, 3.8 percent, which would double its numbers in eighteen years) aims to take more water from the Yarmuk, through a series of dams that will divert 40 percent of the river's flow and induce severe shortages for the farthest downstream country, Israel. Israel asserts that if the largest dam, the Unity Dam, is built, it will consider destroying it.

This, then, is the crunch climax building up in the Jordan River basin. According to a former Israeli minister of agriculture, Meir Ben-Meir, "If the people of the region are not clever enough to discuss a mutual solution to the problem of water scarcity, war is unavoidable." King Hussein of Jordan has declared that water problems will be the only justification for his country to go to war again with Israel. At the same time, of course, the nations in question could view the prospect as a creative challenge.[10] The coming crisis could persuade them that their enlightened self-interest points them in the direction of collaboration rather than confrontation. Could it be that water—the prime resource for all life in the region—will turn out to be the factor that impels them toward a common response to a common problem?

The Euphrates and
Tigris Rivers

There could also be trouble brewing over the Euphrates and Tigris rivers, both of which rise in Turkey before flowing through Syria into Iraq. Thanks to its Tennessee Valley Authority–type Anatolia Project, Turkey plans to construct thirty-eight hydroelectric and irrigation works on the upper reaches of both rivers. While the project will double Turkey's irrigated land and bring the total to more than 7700 square miles (an expanse almost as large as Israel), it could eventually divert almost half of the Euphrates water that now crosses the border into Syria and well over two-thirds of the river's flow into Iraq.[11] Both Syria and Iraq are highly dependent on the rivers for irrigation and electricity.

Of course, much of the water used in Turkey will eventually be returned to the rivers' courses, making it available to downstream consumers. But by then it could have become heavily polluted with irrigation salts and farmland pesticides and fertilizers, plus industrial effluents. The growing salinity of the Euphrates has already put an end to farming in much of the area around the Iraqi city of Basra. And as a measure of the potency of water as a true security threat, Turkey has not hesitated to propose interfering with the Euphrates' flows into Syria in retaliation for Syria's support of Kurdish separatists in Turkey.

Syria, too, entertains ambitious irrigation plans for the Euphrates, which supplies 90 percent of the country's water. A huge dam could eventually divert half the water arriving from Turkey. The downstream consumer, Iraq, would be hit hard, since it receives four-fifths of its water from beyond its borders. In 1975 after Syria built the Ath-Thawrah Dam, the Iraqi government claimed that loss of water threatened the livelihoods of three million farmers. As a result of the dispute, the two countries came to the brink of war.[12] A similar dispute in 1984 almost resulted in hostilities. The new Syrian dam, according to Iraqi estimates, would oblige Iraq to shut down four power plants that supply 40 percent of the country's electricity.

The Nile River

In a different sector of the Middle East, yet another nation is deeply embroiled in a river dispute—Egypt.[13] This carries weighty implications not only for Egypt itself, but for the wider world insofar as Egypt exercises a moderating influence on Arab politics generally. Hardly any other nation is so dependent on a single natural resource as is Egypt on the Nile River for its agriculture.

Two decades ago Egypt could feed itself. Today, and mainly because of population growth and land mismanagement, it faces severe problems on the food front. Food shortages have lead to soaring prices, as a result of which there has been a series of urban riots and other domestic upheavals.[14] In 1988 Egypt had to import nine-tenths of its cereal grains, a proportion that seems set to rise still higher by the year 2000, when the country's present population of 57 million is projected to reach 71 million (soaring to 103 million by 2025). Egypt's foreign debt was $40 billion in 1990, and servicing of nonmilitary debt alone was consuming one-quarter of export earnings. If Egypt cannot reduce its debt burden through increased trade revenues (in recent years its terms of trade have steadily declined), it may not be able to buy all the grain it needs.

What prospect is there that Egypt can grow more of its own food? Due primarily to rapid population growth but also to environmental problems, grain output per person declined between 1970 and 1987 by almost one-fifth.[15] Less than one-twentieth of its land area is cultivated, in a strip averaging six miles wide along the Nile River plus the river's delta. Farming is critically dependent on irrigation, but as far back as 1982 half of all irrigated croplands were suffering some degree of salinization, and all the rest were reckoned at risk.[16] Today about 10 percent of agricultural production is lost annually to decline of soil fertility, and another 8 percent to desert encroachment.

Nor does population planning hold out much hope for Egypt if recent trends persist. There has been little reduction in the population growth rate since 1980, and with plenty of demographic momentum built into present population trends and patterns, even a vigorous birth-control campaign would not show much result for the better part of a generation—by which time the population total is likely to expand by another 35 million people, or almost two-thirds as many as today.

In this tight situation, Egypt faces growing water shortages. Eight successive drought years in the watershed areas of Ethiopia and equatorial Africa reduced the Nile flows by mid-1988 to the lowest level since 1913. Storage water in Lake Nasser looked likely to prove enough for only the 1988 harvest, whereafter Egypt would have had to import a further 15 percent of its food.[17] Fortunately the upstream drought broke in late 1988, and the immediate crisis was relieved. But were the 1980s drought in the upstream catchments to return, as many climatologists anticipate in light of long-standing climatic cycles (possibly to be exacerbated soon by the greenhouse effect),[18] the results would be, according to Mr. Sarwat Fahamy, director of the Nile Water Control Authority, "a calamity."

Moreover, the water shortages would affect not only agriculture. By mid-1988 the low flows into the Aswan Dam's hydropower turbines, which supply 40 percent of Egypt's electricity needs, reduced power output by 20 percent. Even before the drought, the nation was failing to meet a 10 percent annual growth in demand for electricity power. A new drought could easily result in a more serious power cut than in mid-1988, even as high as 60 percent, at a cost of millions of dollars' worth of lost energy each month. Since the deficit would have to be made up with oil, the low-flow problem would all but eliminate Egypt's oil-export revenues—and further restrict the country's ability to buy food elsewhere.

Against this unpromising background, there is emerging an even greater threat to Egypt's water supplies—new claims on the part of upstream nations for a greater share of the Nile's waters.[19] As Vice-President Albert Gore has pointed out,[20] the Nile does not have any more water in it today than it had when Moses was found among the bullrushes, but a lot more people want to drink of its waters (and they will soon become a whole lot more again). The other nations within the river basin— notably Sudan and Ethiopia, together with Uganda, Rwanda, Burundi, Zaire, Kenya, and Tanzania—want to utilize the river to supply their farmlands with irrigation water, rather than relying on rainwater alone. This applies notably in Ethiopia, which controls the Blue Nile tributary and another smaller feeder river, source of four-fifths of the main river's water entering Egypt.

Ethiopia has never joined Egypt (or the other downstream nation, Sudan) in a legal agreement to regulate the parceling of the Nile's waters. On the contrary, it has repeatedly asserted that as a sovereign state it feels at complete liberty to dispose of its natural resources in whatever way it pleases.[21] The Ethiopian government aims to resettle 1.5 million impoverished peasants from the degraded highlands into the nation's fertile southwestern sector. Only 400 square miles have so far been irrigated, out of 11,000 square miles of potential. In order to supply irrigation water for the new settlements, covering 10,400 square miles, Ethiopia plans to eventually divert up to 39 percent of the Blue Nile's waters before they leave its territory.

This prospect alarms Egypt. As long ago as 1980, former Egyptian president Anwar Sadat warned, "If Ethiopia takes any action to block our right to the Nile waters, there will be no alternative for us but to use force because it is a matter of life or death."[22] In 1985 Egyptian foreign minister Butros Ghali was even more forthright: "We depend upon the Nile 100 percent. The next war in our region will be over the waters of the Nile, not over politics. Washington does not take this seriously, because everything for the United States relates to Israel, oil, and the Middle East."[23] And in 1990, during the Gulf War, Ghali spoke up again: "[In] my pessimistic interpretation, I was talking even before the continuing droughts and the problem of desertification. The facts are that in the next few years the demographic explosion in Egypt and the upstream countries will lead to all those countries using more water; and unless we can agree on the management of water resources, we shall have international disputes."

As can be seen from the foregoing, it is certain that were violence to break out again in the Middle East, it would not be over the region's most plentiful resource, oil, but over its scarcest, water. Long after the oil wells run dry, moreover, there will be ever greater competition for water—with all the scope for conflict that entails.[24]

These frictions over rivers in the Middle East highlight the potential for water-related conflicts in other parts of the world. Of 214 major river basins, three-quarters are shared by two nations and one-quarter by three to ten nations.[25] Almost half of Earth's land surface is located within international river basins, supporting 40 percent of the world's population, two-thirds of them in developing nations, which often have less water per citizen than do developed nations. Nearly fifty countries have more than three-quarters of their territory within such areas. Tensions and violence have already erupted in the river basin of the Mekong, shared by Laos, Thailand, Kampuchea, and Vietnam; in that of the Amur, shared by China and the former Soviet Union; in that of the Parena, shared by Brazil and Argentina; in that of the Lauca, shared by Bolivia and Chile; and in that of the Mejerdah, shared by Tunisia and

Libya.[26] A long-standing dispute over the Shatt el Arab on the border between Iraq and Iran contributed to the outbreak of the 1980s war.

Water Anemia

To clarify our understanding of water shortages, let us consider the analysis of a leading hydrologist, Dr. Malin Falkenmark of the Swedish Natural Science Research Council.[27] She focuses on the number of people who can survive off what is technically known as a "flow unit," or one million cubic meters of water per year. Experience in developed countries, usually possessing a sophisticated capacity for water management, shows that when the number exceeds 2000 persons, meaning less than 500 cubic meters of water each, there are likely to be endemic water shortages. Or, to put it in the jargon of the trade, "the hydraulic density of population causes water stress." In developing countries, where water management is likely to be less advanced and rainfall is often concentrated in a limited part of the year, the water-stress level generally occurs as early as 1000 persons per flow unit.

In Jordan the figure in question is already 3000 persons per flow unit, a dire situation indeed. In Libya it is almost 3500 persons, and in Saudi Arabia almost 4000 persons. But in these latter two countries it is acceptable since they have the oil wealth to deploy the most modern technologies to recycle water time and again, or to buy food overseas. In Britain, France, and Italy, the flow-unit number is only 350: zero problem.

In much of Africa, however, there is already a prime problem of water scarcity in several countries. In many of the rest, the problem is waiting down the road. (When it rains in New York, people make a face; in Africa they dance in the streets.) It lies not only with meager amounts of rainfall in total. It relates also to rainfall distribution around the year. Two-thirds of the region receives more than half its annual rainfall in just three months, restricting the length of the crop growing season. In turn this means that in roughly half the continent, drought imposes a critical limit on agriculture except in those few localities where irrigation makes up the difference.

As a result, large numbers of Africans suffer from water scarcity (Table 3.1). By way of comparison, they experience deficits on a scale equivalent at least to water problems now afflicting the lower Colorado basin in the United States. By the year 2000, all five North African countries and six out of seven East African countries are projected to be suffering "water stress." These, plus communities in water-short areas of other countries, will number 350 million people, or almost half the continent's total. By the year 2015, they will be joined by another ten countries. All in all, these countries' populations are projected to number 1.1 billion people, or well over two-thirds of the continent's total.

Much of the problem stems from population growth. Africa is not only the driest region of the developing world, it also features the highest birth rates. In most of the continent human numbers are projected to increase threefold by the year 2035 or shortly thereafter. If we count backwards from the "water barrier" figure mentioned above, namely, 2000 persons per flow unit, and divide by three (to allow for the tripling of population), we find several countries already above the 670-person level, notably Egypt at 690, Morocco at 710, and Kenya at 910—with several more poised to follow suit. In these countries there will likely be acute water shortages well before the year 2035. A few countries will find themselves in difficulty even within the near future. During the last two decades of this century, Egypt's water supply per person is projected to shrink by one-third, Nigeria's by two-fifths, and Kenya's by one-half.[28]

Nor is the problem a straightforward affair of too many people and too little water. It is intensified by a number of

TABLE 3.1
WATER SCARCITY INDEX
IN SELECTED AFRICAN COUNTRIES

Country	Population (millions)			Level of Water Competition (people per million cubic meters per year; water stress induced by roughly 1000 people at most)		
	1982	2000	2025	1982	2000	2025
Algeria	20	35	57	650	1100	1900
Burundi	5	7	11	1200	1900	3100
Egypt	44	69	97	460	690	1000
Kenya	18	39	83	480	1000	2200
Malawi	7	12	23	730	1300	2600
Morocco	22	36	60	680	1100	1900
Nigeria	82	162	338	270	530	1100
Rwanda	5	11	22	810	1680	3500
Somalia	5	7	13	430	600	1100
Tanzania	19	39	84	250	520	1100
Tunisia	7	10	14	1500	2100	3000
Zimbabwe	8	15	33	350	660	1400

Source: M. Falkenmark, "Middle East Hydropolitics: Water Scarcity and Conflicts in the Middle East," *Ambio,* 18 (1989):350–352.

further factors, each of which tends to reinforce the others as well. In many countries the good-rainfall areas have already been taken for agriculture. Featuring too many farmers as a result of population growth, these overburdened lands lose top-soil and plant nutrients; without sizable technical inputs, some of them even fail to support as many people as before. Growing numbers of farmers are obliged to head for new farmlands. The same factor of population growth also means that there are still more people seeking to live off the countryside. More people looking for more land means that drier areas are brought under the plow or hoe. As well as loss of forests and other water-conserving vegetation, the process seems to be triggering a desiccatory effect across extensive areas, resulting in less rainfall again.[29]

The overall upshot, to cite Dr. Falkenmark again, is a "risk spiral" where one factor is worsened by others, whereupon it supplies its own adverse impact in turn but with increased force. The syndrome of multiple factors compounding each other's impacts receives scant attention from analysts, even though it could well turn out to be the most important of all. Put them together and these factors add up to a somber prospect: already dry Africa could shortly experience a "drying out." As demand for water soars faster than ever, supplies could eventually decline faster than ever.

This risk spiral will engender severe repercussions for African countries' capacity to feed themselves, and hence for their governments to maintain stable regimes and security. Of the world's 150-plus river basins shared by two nations, 57 are in Africa, a highly disproportionate share in relation to area (Table 3.2). The Niger River flows through ten countries, the Nile and Zaire rivers through nine, and the Zambezi River through eight. Nor will water shortages affect relations only between countries. We can learn from experience outside Africa. Already there have been outbursts of violence over water within countries, too. In India, for instance, constant

Table 3.2
WATER AND
INTERNATIONAL RIVER BASINS

Country	1000 Cubic Meters per Person per Year* (1988)	Territory in International River Basins (%)
Afghanistan	2.5	91
Iraq	1.9	83
Sudan	1.2	81
Ethiopia	2.3	80
Bulgaria	2.0	79
Pakistan	2.7	75
Syria	0.6	72
Botswana	0.8	68
India	2.3	30

*When water availability is only 2000 cubic meters per person per year, it is considered to represent a situation of serious water shortages.

Source: P.H. Gleick, "Climate Change and International Politics: Problems Facing Developing Countries," Ambio, 18(6, 1989):333–339.

clashes have erupted in Punjab, where Sikh nationalists claim too much of their water has been diverted to the Hindu states of Haryana and Rajasthan. There is acute competition for water in the vast expanse of fertile farmlands that make up the north China Plain, a region producing one-quarter of China's grain. Since Beijing's demand for industrial and domestic water is expected to expand by 50 percent during the 1990s (its water table is dropping at a remarkable six feet per year), farmers in the environs could be deprived of as much as two-fifths of their cropland water.[30]

Water shortages are not confined to the developing world. In the central Asian republics of the former Soviet Union, surging demand for irrigation water since 1950 has caused the Aral Sea to shrink by almost half, leaving behind almost 10,000 square miles of human-made salt desert. Once the world's fourth largest freshwater lake, the Aral Sea is likely to decline to a few residual brine lakes within another few decades. Local scientists view it as an environmental catastrophe to rival Chernobyl. The demise of the Aral will act to worsen water shortages in an extensive sector of Central Asia, contributing to political tensions.

Similarly, in the United States, one-fifth of irrigated lands are watered by excessive pumping of groundwater. The Ogallala Aquifer underlying several western states is the largest aquifer in the world; it is being depleted at a rate thousands of times that of natural replenishment. Such is the pressure for water in the western states that the Colorado River is becoming too salty for most purposes by the time it flows across the border into Mexico. Already it has become a source of international friction.

Water and Health

There is a further dimension to water shortages: health. Four out of five deaths in the developing world are due to water-related diseases such as malaria, cholera, yellow fever, schistosomiasis, and especially diarrhea.[31] This serves not only to intensify demand for water by fast-growing populations, but reinforces the absolute poverty endured by the "bottom billion," since sick people are far less able to grow food and otherwise support themselves. Recall, for instance, the devastating impact of the 1991 cholera epidemic in Peru, now spreading to other countries, and due primarily to lack of clean water. Invariably the epidemic has hit hardest at the poorest.

While this ill-health dimension of the water issue would be tough enough in itself, it links up with other issues. Insofar as 35,000 of readily avoidable deaths in the developing world each day are children, there is less motivation for parents to engage in family planning while they see all too many of their offspring succumbing to disease. Equally important though less readily recognized, impoverished people readily become desperate people, all the more likely to support guerrilla movements, if not to engage in direct resistance to governments.

The number of developing-world people who still lack safe drinking water is 1.2 billion, or well over one-fifth of humankind. The number who lack adequate sanitation facilities, preferably water-based facilities, is 1.8 billion, or every third person. While the first number has declined a bit since 1980, the second has grown. To supply the water needs of four-fifths of developing-world people would cost around $15 billion a year for ten years, mainly through making more efficient use of water stocks. (To put this figure in perspective, note that waterborne diseases now cost India alone $600 million a year through additional health bills and lost worker output; in the developing world as a whole, waterborne diseases cause at least 500 million working days to be lost each year, and if each day is worth $10 on average, that works out to $5 billion each year.) Simple technologies such as hand pumps would do much of the job. Yet governments and international agencies have come up with only one-third of the target budget. Hardly a single developing country is achieving its water-supply goal.

The annual cost is not great in further comparative terms. It is equivalent to just five days of military spending by the community of nations. While we persist in saying effectively that we cannot afford to do the job, the world—a world with many more people arriving—is growing ever more thirsty.

To summarize, let me note three points. First, water shortages are a severe problem for scores of countries, a problem that is set to grow worse fast. Second, there seems to be no ready solution short of massive water-management programs with hefty price tags that apparently remain beyond the capacity of governments and international agencies alike. Even that would hardly provide a long-term solution for those many countries that are going to run up against "water barriers" if only because of population growth. Third, water shortages have already triggered a number of disputes and conflicts, and these confrontations show every prospect of growing more numerous and violent. It is hardly overstating the case to assert that water could soon become a prime cause of what is likely to be a salient phenomenon of our future world—resource wars.

Remarkably enough, however, the issue is scarcely recognized as a potent source of conflict. In fact, water shortages themselves have received hardly any play to date. The World Commission on Environment and Development offered only a passing mention in its 1987 report[32]; nor did the issue figure much at the 1992 Rio Conference on Environment and Development. During the 1990s, however, water shortages could become as significant as food shortages were during the 1970s and energy shortages during the 1980s. Most countries with too little food are countries with too little water. Already two billion people live in areas with chronic water shortages, and their number is rising rapidly. After long remaining a sleeper issue, water could shortly emerge as a top-line item on international agendas if only for security reasons.[33]

Scenarios

A downside scenario foresees that water shortages in the Middle East will lead to "water piracy" on the part of a nation that views its crucial interests as threatened by declining water flows for agriculture. Other nations rush to follow the precedent, with piracy, counter-piracy, and terrorists attacking dams, pipelines, pumping stations, and water-treatment plants. These skirmishes escalate into localized hostilities that eventually lead to the outbreak of war. Egypt, feeling driven to desperation by Ethiopia's diversion of the Nile's waters, launches a preemptive strike against dams being built on upstream tributaries. Ethiopia, in retaliation, bombs the Aswan Dam.

Further afield, nations of the Indian subcontinent trade "water ultimatums" with ever greater vigor. Growing hunger in northern China, where water supplies have been tightening for decades, culminates in a "water uprising." Still other nations suspect that global warming will disrupt their rainfall patterns to an extent that will leave them with less water precisely at a time when they need much more to supply their burgeoning populations. In particular, they note a forecast that warns they may lose more than 50,000 square miles of irrigated lands and a still larger amount of rainfed croplands in a greenhouse-affected world, requiring compensatory measures costing between $145 billion and $300 billion.[34] They start to engage in "beggar your downstream neighbors" tactics by an orgy of dam building and other measures to safeguard "their" water. Confrontations proliferate, followed by conflicts.

By contrast, an upbeat scenario anticipates that water could come to be seen as a common challenge that warrants common strategies: The Middle East nations slowly recognize they share a region where one of their most vital resources does not acknowledge international frontiers. Without water to sustain human communities, disputes over parched territories are finally seen to serve limited purpose: what point is there in "victory" over land that cannot support agriculture? Old hurts and new hatreds are both overtaken by a "let's start afresh" spirit that is inspired by environmental imperatives. Eventually the region's nations find themselves obliged to agree that confrontation over water must give way to cooperation over water. By force of environmental circumstance, they are induced to dismiss military force as a solution.

Following the peace conference begun in 1991 when water figured high on the security agenda, they accept the enlightened leadership of the United States, one of whose State Department officials, Richard Armitage, has long warned his own national leaders that "[i]t is time for the United States to acknowledge that the Middle East water crisis is worsening and adding an extra dimension to prospective war scenarios."[35] All parties start on the track toward a regional settlement. As a result, peace releases a sizable flow of funds from the U.S. government, the World Bank, and other agencies, supplied to boost economies that have been drained for decades by preparations for war.

The Middle East breakthrough persuades other nations to consider the same path. The worldwide Water Security Convention of 1995 soon attracts sufficient signatories to establish water collaboration as a norm of international relations. This departure builds up an international expectation that water stocks shall be viewed as "everybody's business," and a series of pioneering precedents quickly reinforces the concept as a basic mode of international behavior.

Could we foresee, then, that water will one day become the lubricant to smooth the way toward a more stable and secure world? Or, to shift the metaphor, could water eventually serve to douse the sparks that would otherwise ignite the flames of war?

4

Ethiopia

The greatest problems facing the world today, beside the threat of nuclear war, are deterioration of the environment and its link with poverty and the bleak prospects for development.

—Javier Perez DeCuellar,
Secretary-General
of the United Nations, 1988

I first visited Ethiopia in 1972. I was struck by the extreme beauty of the highlands, where the bulk of the populace lives. I was struck, too, by the extreme poverty of most people, and by the extreme devastation of the landscapes. I could well believe that while half of the highlands had been forested as recently as 1950, it had fallen to one-twentieth as much as a result of excessive fuelwood cutting and the need to hack out new croplands for the burgeoning population. I wondered how long the situation could persist before it broke apart.

But all seemed serene enough in 1972. On every side were people going about their peaceful ways, impoverished though they were. Ethiopians are different from other Africans. With a 4000-year history and an early ruler who was the alleged son of Solomon and the Queen of Sheba, they are thought to have originated from outside Africa. Their religious background is centered in part on Coptic traditions, as witness the equal-armed cross that is prominent in much Ethiopian art. Perhaps most important of all, and by contrast with virtually all the rest of Africa, Ethiopia has never been colonized except for a brief incursion by the Italians. This is reflected in the demeanor of many Ethiopians: theirs is an independent tradition, their culture has been little modified by outside influences, and they know it. Cut off in their highlands home, they seem to have been living in a semi-lost and distinctively different world.

This facet of Ethiopian life has always registered with me whenever I have set foot in the country. How like them, I have mused, that at the Rome Olympics of 1960 they should produce an athlete, unknown till then, who led home the marathon field while running in bare feet, and who on completing the course lay down in the stadium infield to engage in vigorous calisthenics as the also-rans stumbled to the finish. He went back to Ethiopia and was hardly heard of again until he turned up at the next Olympics and repeated his feat. Tough people, I have reflected: they must be to survive in the devastated environments of modern-day Ethiopia.

Little did I suppose in 1972 that within two years the smoke of burning forests would give way to the smoke of revolution. Ten years later I contributed to an environmental assessment of the late 1970s war between Ethiopia and Somalia, prepared for a peace commission under the Organization of African Unity. The more I analyzed the "big-picture" background to the war, the more the message became plain. Environmental problems had played a crucial though covert role in setting off the war.

Today I feel that Ethiopia serves as a prime example of environmental insecurity. During the 1960s there was widespread soil erosion in the country's highlands. This was due primarily to deforestation, though due also to rudimentary agricultural practices, inequitable land-tenure systems, and the pressures generated by a population that increased from 18 million in 1950 to 25 million in 1970. As a result of the soil erosion, there was a decline in farmland fertility and a hefty falloff in agriculture, followed by food shortages and spiraling prices. It all culminated in riots in Ethiopia's cities, eventually precipitating the overthrow of Emperor Haile Selassie in 1974.[1] This was the first time a government had been ousted for primarily environmental reasons, but as we shall see in this book, it has certainly not been the last.

The Ogaden War

The new regime under President Mengistu—one of the worst tyrants to afflict Africa—did not move vigorously enough to restore agriculture, despite some commendable efforts at cropland terracing for soil conservation. So throngs of impoverished peasants, finding they could no longer make a living in the highlands, started to stream toward the country's lowlands, notably the Ogaden zone straddling the border with Somalia. In Somalia, too, steadily increasing human numbers, together with inefficient agricultural practices, had led to much overtaxing of traditional farmlands.[2] For these reasons (plus some ethnic complications that had generated a long-standing dispute between the two countries), there was a migration toward the Ogaden from the Somali side as well.[3] The Somalis felt threatened by this turn of events; they were also worried the Ethiopian migrants might divert river water for farmlands—and almost all of Somalia's rivers rise in Ethiopia.[4] In response to the threats, the Somalis made numerous bellicose noises. In 1977 the two countries went to war.[5]

If this had been a local punch-up between two of the poorest countries on Earth, that would have been regrettable enough, while of scant consequence to the rest of the world. But it occurred in the Horn of Africa, a jutting-out sector of land that borders onto the Red Sea and the Indian Ocean, both of them zones that were growing prominent in geopolitics. So the war mattered a great deal to outsiders, especially to the United States, if only because of the strategic oil-tanker lanes from the Persian Gulf to the industrialized Western world.[6] Since the new Marxist regime in Ethiopia was already supported by the Soviet Union, the United States came in on the side of Somalia.

Neither superpower provided anything except armaments. But they did that in a big way. Thanks to external munificence, Ethiopia between 1977 and 1980 spent an average of $225 million a year on military activities, rising to $378 million in 1981 (some, too, for activities in the rebel provinces of Tigre and Eritrea as well as in the Ogaden); Somalia spent almost as much.[7] If one-quarter of the sums consumed by the Ogaden hostilities had been assigned beforehand to safeguarding topsoil, tree cover, and associated factors of the agricultural-resource base in traditional farmlands of Ethiopia, the migration toward the Ogaden would have been far less likely.[8] Ironically a similar amount of finance was required from the outside world in relief measures alone to counter Ethiopia's 1985 famine—a famine largely induced by environmental rundown.

In still other ways, soil erosion had been levying a stiff price on Ethiopia's economy. Loss of plant nutrients in the soil had been cutting food output by at least one million tons per year, equivalent to two-thirds of all relief food shipped to the country in 1985.[9] Moreover, as trees and sources of fuelwood disappeared too, Ethiopian farmers turned to burning cattle dung and crop residues. So much of these residues—between 60 and

90 percent in some areas—were being used as fuel instead of fertilizer that there was a further decline in agricultural output, worth $600 million per year by the mid-1980s[10]—a figure to throw further light on the cost of environmental rehabilitation, $500 million over ten years.

Enfamished Ethiopia

Today famine has become a way of life in Ethiopia.[11] The natural-resource base underpinning agriculture—soil, vegetation, water—has been so severely and widely depleted that the nation can no longer perform its most basic and vital function, that of supplying its citizens with enough food.[12] In consequence, Ethiopia scarcely operates any longer as an independent nation in the usual sense of the term. Two-thirds of food supplies are provided by outside agencies rather than by the nation itself. As we shall see in the case of many other countries of sub-Saharan Africa (Chapter 5), Bangladesh (Chapter 7), and El Salvador (Chapter 8), Ethiopia is increasingly unable to fashion its own future. Many critical decisions for the country are taken by other governments, not all of which have at heart the best long-term interests of the country as understood by Ethiopians.

The 1985 famine-relief effort to help Ethiopia was the largest ever provided to an individual country, supplying as much as one-fifth of food required. Several hundred thousand people died, but without outside help the total would have been many millions. Yet the situation keeps on going from bad to worse, with a decline in food production averaging 1 percent per year even while the population continues to expand by almost 3 percent per year.[13] Ethiopia must surely be facing a future where famines become more frequent and drastic.

Nor are the portents hopeful short of super-scale efforts. A study appraising the prospect for the next fifty years, undertaken by Professor Hans Hurni and his colleagues of the Geography Institute at Bern University, Switzerland,[14] postulates a scenario where the population grows as projected to 200 million people by the year 2040, while environmental safeguards together with development efforts continue on their recent dismal track. Grazing land will become scarce in the extreme within the next two decades at most, and there will be livestock crises in over half the country as soon as the year 2000. Forests will virtually disappear, meaning that soil erosion will increase and water systems will decline, and more livestock manure will be used for fuel rather than fertilizer; deforestation has been levying environmental costs of between 6 and 9 percent of GNP.[15] Even with mechanization of one-third of farmlands, agriculture will provide the basic needs of no more than 100 million people. To relieve the crisis, agriculture improvement measures will have to be doubled, soil conservation and other environmental programs increased fourfold, livestock upgrading expanded sixfold, and population-planning activities increased sufficiently to limit the ultimate population to only 100 million people at most.

As a result of the imbalance between population and food, plus associated political upheavals, there are at least 7 million people threatened by starvation in Ethiopia today, and a total of 20 million people are chronically undernourished.[16] As a further result, there are at least 3 million displaced people within Ethiopia, and another half million in refugee camps in Sudan where they are viewed as environmental rather than political refugees.[17]

In short, environmental breakdown and food shortages have become endemic in Ethiopia.[18] It is now apparent that the mid-1980s drought was no more than a triggering factor for a crisis that had been building up through the pressures of environmental ruin and population growth. While we can hardly assert that population growth has been the prime or direct cause of the recent turmoil in Ethiopia, it has certainly helped to exacerbate the problem insofar as the population has tripled since 1950. A speculative appraisal of what could have been achieved by population regulation shows that if Ethiopia had established a strong family-planning program from 1970 onwards, and if the program had achieved only moderate success, it might have prevented 1.7 million births by 1985 at a cost of around $170 million. Ironically, the emergency food rushed to Ethiopia during the 1985 famine supplied enough food to sustain 1.7 million people, also at a cost of around $170 million.[19]

Northeastern Africa features other flashpoints of conflict in those countries where fast-growing populations are pressing against the limits of their natural-resource base. In the wake of its recent revolution and civil war, Somalia hardly functions as a nation anymore, so ravaged has it become after years of fighting and decades of overburdening of its environments nationwide.[20] Consider, too, Sudan, and some root causes of the 1985 coup that toppled the Nimeiry regime. Four problems appear to have been at work.[21] First was the civil war in the south, provoked by fears that water that "belonged" to the southerners was being diverted to the northerners. Second was a 500 percent rise in the price of basic food staples as a result of drought and associated problems of declining agriculture. Third was a severe shortage of fuelwood in the north, after local supplies had been sorely depleted and the main supplementary supply from the south had been cut off, precipitating a demand for kerosene and gasoline with price rises of more than 300 percent. Fourth was the mass migration into the Khartoum area from the Kordofan and Darfur provinces, where gross deterioration of grazing lands had been aggravated by drought. All of these factors had a strong environmental component—and all were accentuated by the pressures of a population growing at 2.8 percent per year.

Back to Ethiopia. Our brief review has made plain that environmental factors, while rarely cited in news headlines, have contributed much to today's situation—an ultimate debacle which has heralded the breakup of Ethiopia as a nation. Devastation of the natural-resource base, accentuated by almost twenty years of civil war, has left the economy in shreds. Disaffected by the indifference of two successive dictatorships, the northern province of Eritrea wants to go it alone, depriving the rest of the country of the sea outlet of Massawa, which has been handling almost three-quarters of overseas trade. Two dozen other ethnic groups with their liberation forces likewise want out to some degree or another.

The Wider Impact

The disintegration of Ethiopia carries perils that are not confined to that benighted country. It promises to spill over far and wide, spreading instability in a good many other nations of sub-Saharan Africa. Following the departure of the colonial powers roughly thirty years ago, African nations found themselves left with borders that harked back to absurd arrangements from late in the nineteenth century. Many borders today divide ethnic communities that give their allegiance to local tribal unity rather than to national entities with remote governments. Queen Victoria once cast around for a birthday present to give to her cousin German Kaiser Wilhelm; she decided to offer the transfer of Kilimanjaro from Kenya to Tanganyika, thus splitting one part of a tribe from its brethren on the other side of the new border.

African leaders have long feared ethnic clamors for realignment of their countries' borders. Secessionist urges led to civil wars in Zaire, Nigeria, and Sudan, and to simmering revolt in a dozen other countries. Politicians rightly suspect if one country allows its disaffected factions to break free, that will spring loose a Pandora's box of similar demands that would challenge their status as national leaders. For three decades the Organization of African Unity has tried to keep the lid on by proclaiming the sanctity of borders, even though this has frustrated local unity throughout the region. If Ethiopia allows itself to be dismantled by liberation groups, the precedent will encourage many other such groups to raise the banner of revolt.

Thus the travails of Ethiopia seem set to disrupt lands way beyond its borders. An environmental debacle in one country reaches like an infection to undermine the body politic of an entire region. The original cancer can metastasize one knows not where. Of course, there is no direct cause-and-effect relationship. The linkages are covert and diffuse, with many other factors mixed in. But however difficult to discern, the linkages are real, numerous, and significant. Environmental issues and their repercussions can extend far and wide, albeit through indirect mechanisms. They are like stones dropped in a pond, sending out ripples that combine with ripples from other stones. In modern-day Africa, they recognize no frontiers—however little that fact is heeded by politicians with their fixation on frontiers. A remote corner of Africa now threatens security way beyond its horizons, beyond the political horizons, too, of traditionalist leaders with their purview limited to a world that has disappeared as much as the former colonial world.

There is a further dimension to Ethiopia's environmental troubles, this one extending way outside Africa. The country possesses an exceptional array of wildlife species. Plant species alone, never mind animal species, total 6000, or one-third as many as in the seven-times-larger United States. One-tenth of them are endemic, that is, found nowhere else on Earth. Most of them are located in forests; and as the forests decline, so do their species. How many plants have disappeared for good in Ethiopia is anyone's guess. Probably scores already.

Fortunately one plant that survived long enough to supply its economic services to the wider world is a wild relative of barley, harboring genetic resistance to yellow dwarf disease. The germplasm has been bred into commercial strains of barley, supplying benefits to American farmers alone worth $150 million per year.²² Let's remember that the next time we reach for our evening beer. Still other countries enjoy spinoff services from Ethiopia's wild plants. Ethiopia is the native home of coffee, and its forests support a good number of wild relatives which supply genetic resistance to various coffee crop blights. Several countries grow coffee, but none more than Brazil, where coffee is a major export. Ironically Brazil has been prominent in proclaiming that any country's forests, notably its own in Amazonia, are strictly the business of the "owner" country, and the same for the genetic resources therein. Yet Brazil has been quietly urging Ethiopia to take better care of its forests, especially now that they are reduced to mere fragments of what they once were and are still declining. If Ethiopia's forests and their gene stocks finally disappear, we may face the $3 cup of coffee. To paraphrase John Donne, no man, and no country either, is an island.

Scenarios

A downside scenario sees the secessionist elements in Ethiopia becoming dominant and the country splits into several ministates, none of them viable as independent units. The outside world, both within Africa and farther afield, quickly loses interest. It believes there is little strategic value at stake in the Horn of Africa, so why bother with a group of separatists whose collective GNP is only $7 billion, or not even one-tenth that of New York or London? The oil-rich Arabs contribute nothing since Ethiopia is not Muslim. Lacking much in the way of external support except from a few humanitarian groups such as Oxfam, the new entities are unable to restore their environments and rebuild their economies. The last fragments of forest disappear (hundreds of plant species with them), grasslands are overburdened more than ever, and soil erosion grows worse on every side. Agriculture falls into such disrepair that it scarcely warrants the name. Famine and starvation become the order of the day, and the death toll climbs year after year. But the appalling mortality still does not cancel out the birth rate as impoverished peasants consider their main salvation lies in producing more offspring than ever, thus compounding the destructive impact of too many people struggling to survive off a devastated resource base.

The 1996 hunger disaster results in a human tragedy surpassing anything of its sort in modern times—more than 20 million starving, as compared with 8 million in 1990. The United Nations speaks of one million dead from starvation and malnutrition-related diseases, but they are replaced by eight months of population growth as the "old Ethiopia" tops 60 million inhabitants, three times as many as in 1950. More millions flee into neighboring Sudan, but this country, too, is suffering unprecedented problems of similar kind and eventually its host-country patience runs out. It seals its border with the "Ethiopia zone." The famished throngs try to fight their way into Sudan, leading to outbreak of a border war that promises to sputter on indefinitely.

The world remains indifferent. Whereas Ethiopia's plight could once fill London's Wembley Stadium for Bob Geldof's rock concert and hold the attention of one-fifth of humankind through television, springing loose $65 million from concerned citizens around the world, the international community has now wearied of a continuing calamity that looks as if it is beyond hope. Experts forecast that by the year 2000 the area will suffer as much mortality as all sub-Saharan Africa in 1990, a time when the situation could still have been saved. Nobody seems to listen. But a few local activists, supported by the Libyan malcontents who overthrew Colonel Khaddafi as too easygoing, seize the opportunity to "take an initiative" by resorting to terrorism against the rich nations, using the latest micro-weapons technology to drive their point home. Like many terrorists, their main demand is a hearing.

An upbeat scenario foresees the would-be separatists soon learn that they have far more to gain together than to lose separately. They engage in a creeping confederalism that steadily pulls them together again. They mobilize the community-level spirit that inspired the land terracing and soil conservation programs in a few localities during the 1980s, and they engineer broad-scale tree planting right across the highlands, reforesting 20,000 square miles in short order. Agroforestry efforts restore farmland fertility and river flows for new irrigation networks. In 1997 the country achieves record grain harvests, and for the first time in three decades it can aim for eventual self-sufficiency in food. A "deep democracy" government is hailed by the world for its success in boot-straps progress by those with no boots. The citizens become sufficiently confident that the government truly cares about them that they respond to a radical call for family planning by surpassing all targets forthwith.

Meantime the outside world acknowledges its shameful disregard for Ethiopia during its time of travail—spurred on perhaps by the discovery that a disastrous leaf blight among cabbage, broccoli, cauliflower, and other brassica crops in Europe has been beaten by genetic resistance in a wild relative found in the ancestral home of brassica plants, Ethiopia. A new American president tells of his reaction at viewing the devastated grasslands in a yet-to-be-rehabilitated zone, and concedes that bigger-picture factors got lost from view in the long grass of Washington. He offers economic recovery funding, to which the Ethiopian president responds that it would be gratefully accepted as a recovery reinforcer, but that recovery itself is already a latter-day saga in Ethiopia's 4000-year history.

5

Sub-Saharan Africa

Africa may be the major area of conflict in the twenty-first century. We have the possibility of real, serious catastrophe—in terms of human suffering, in terms of violent conflict, and in terms of a retrogression in development. But these factors can be reversed if the world can collectively take care of resource distribution, population regulation, and control of the environment. Otherwise Africa will become a destabilizing factor in the equation of international peace and security.

—General Olusegun Obasanjo,
Former Nigerian Head of State

I have lived more than half my adult life in sub-Saharan Africa. Mostly based in Kenya, I have traveled around the region repeatedly, visiting virtually all its countries. For many years I was pleased and proud to be a citizen of Kenya, a land which I still look upon as the closest to "home" that I have ever known. When I arrived in Nairobi in 1958, I found a country as emergent as any, and for most of its time since independence Kenya has been something of a success story. Alas, Kenya is now into difficult times, as are most countries of the region. What hope is there that sub-Saharan Africa can look forward to a safari into a brighter future? The region suffers much environmental devastation. It features a higher proportion of land losing soil fertility than does Asia or Latin America. Already the most impoverished region in the world, it is blighted by a persistent spread of poverty. Discounting South Africa, its combined economies are no larger than that of Belgium, a country with only one-sixtieth as many people. Today's per-capita income of around $300 per annum is less in real terms than in 1960; almost two out of three people subsist in absolute poverty. Of the world's thirty-six poorest countries, twenty-nine are here.[1] Most people receive less food than when their countries raised the flag of nationhood thirty years ago. In all these respects, sub-Saharan Africa is a good way behind the other two main developing regions, Asia and Latin America, and it is falling still further behind. It is becoming the third world of the Third World.

At the same time, sub-Saharan Africa is racked with political turmoil and violence. There have been more than 200 coups or attempted coups since 1950. We have seen (Chapter 4) how Ethiopia has endured prolonged war; the same goes for several other countries, notably Sudan, Chad, Angola, and Mozambique. Each of these countries also suffers widespread environmental travails in conjunction with wall-to-wall poverty. Can it be coincidence?

The region also features a higher rate of population growth than the other two main regions, and in a number of its countries the rate is still climbing (Table 5.1). Around one-third of the increase in humanity's numbers during the foreseeable future is projected to occur in the sub-Saharan region, which in turn is projected to feature a fivefold increase in numbers. Yet there are signs that already the region suffers from population pressures, which worsen if not cause the environmental predicament—especially as concerns the agricultural-resource base.[2]

TABLE 5.1 SUB-SAHARAN AFRICA: SELECTED BASIC INDICATORS

Country	Population Mid-1992 (millions)	Population Growth Rate 1992 (%)	Projected Population (millions) Year 2000	Projected Population (millions) Year 2025	Projected Final Population (millions)	Per-Capita GNP 1990, U.S. $ (and average annual growth rate 1965–90, %)	Average Annual Growth of Agricultural Output 1980–90 (%)	Foreign Debt as % of GNP 1990
Western Africa								
Ghana	16.0	3.2	20	35	62	390 (−1.4)	1.0	57
Ivory Coast	13.0	3.6	17	39	64	730 (0.5)	1.0	205
Nigeria	90.1	3.0	153	216	453	370 (0.1)	3.3	111
Eastern Africa								
Ethiopia	54.3	2.8	71	140	420	120 (−0.2)	−0.1	54
Kenya	26.2	3.7	34	62	125	370 (1.9)	3.3	81
Madagascar	11.9	3.2	15	32	46	230 (−1.9)	2.4	134
Mozambique	16.6	2.7	21	36	97	80 (NA)*	1.3	385
Sudan	26.5	3.1	33	57	102	NA (NA)	NA	NA
Tanzania	27.4	3.5	33	78	146	120 (−0.2)	4.1	282
Uganda	17.5	3.7	23	50	92	220 (−2.4)	2.5	92
Middle Africa								
Cameroon	12.7	3.2	16	36	53	940 (3.0)	1.6	57
Zaire	37.9	3.1	50	98	172	230 (−2.2)	2.5	141

*NA = not available.

At least half of all people receive less than 90 percent of the minimum calories needed to support an active working life; that is, they are chronically undernourished. Worse, 50 million are simply starving.[3] All but four of the forty-six countries regularly require food aid, and even so fourteen of them do not receive enough to meet basic needs.[4]

Yet today's population of 550 million is projected to reach 668 million (21 percent more) by the year 2000 and 1.229 billion (123 percent more) by 2025.[5] This is an ultra-rapid rate of growth that unfortunately operates in conjunction with adverse climatic conditions such as have been the norm in much of the region for the past two decades. If these adverse conditions persist, they could well start to be aggravated by the climatic upsets of global warming,[6] and the famished throngs that totaled 30 million in 1985 could well increase to 130 million by the year 2000.[7] This means that the proportion of starving people would expand from less than one in fourteen to almost one in five.

All in all, sub-Saharan Africa serves as a prime example of an "adverse-outlook" scenario.[8] If the region continues to sink into widespread depravation, will this not supply conditions for political turmoil within dozens of countries—and thus offer scope for general instability and myriad threats to security?

How Are Africans to Feed Themselves?

Since 1960 the region has been growing hungrier in absolute as well as relative terms. Worse, the World Bank estimates that agricultural production is unlikely to grow at more than 2.5 percent per year for at least the next two decades, even while population growth remains at 3 percent or more per year.[9] As a result, food output per head, which has declined by 20 percent since 1970, is scheduled to decline by a further 30 percent.[10] As a further result, the food gap will grow from 12 million tons in 1993 to 250 million tons in 2020. Even if food production could be boosted from an average annual growth of 2 percent to 4 percent, while population growth remained at 3 percent, the gap would still increase to 110 million tons. Only if food production could be boosted to 4 percent and family size were reduced by half would the gap be brought below the level of 1990.[11]

Doubling the annual growth rate of food production would be a formidable task. Most good-quality lands have already been taken for crop growing. This is the driest of Earth's major regions; rainfall is low and unpredictable. Four-fifths of soils are readily erodible; soil erosion has increased threefold during the past three decades,[12] and desertification already affects more than one-quarter of the region with the agricultural potential lost for decades.[13] Given recent trends, it is hard to see how the pressures of huge numbers of subsistence peasants will not deplete the land's productivity still further—rather than enabling it to become twice as productive.

In terms of individual countries, as far back as 1975 there were fourteen countries, accounting for one-third of the region's land area and half of its population, that featured populaces already too large to grow enough food for themselves assuming continuation of subsistence farming methods (an assumption that has been borne out in most of the region). Even if agricultural practices were to be upgraded with intermediate technology to reach commercial levels, half of these countries would still not manage to feed themselves from their own land by the year 2000.[14] Among these nations are Senegal, Mali, Burkina Faso, Sierra Leone, Togo, Benin, Malawi, and Swaziland, none of which is expected to be generating enough exports to allow purchase of sufficient food stocks abroad. A further lengthy list of countries will be unable to feed even half their projected year 2000 populations from their own land: Niger, Nigeria, Ethiopia, Somalia, Kenya, Uganda, Rwanda,

Burundi, and Lesotho. Of these, only Niger, Nigeria, and Kenya are likely to be exporting enough to enable them to purchase much food elsewhere.[15]

Note the particular case of Kenya. Today's total of 27 million people is projected to expand to an eventual 125 million by the time it levels out at zero growth. Yet even if Kenya were to practice high-tech agriculture, it could not feed more than 60 million people from its own lands (whose soil fertility is widely declining anyway). The country's prospects of exporting enough to buy food enough for its fast-growing populace are doubtful. Kenya's terms of trade have been stagnant or declining for several years. Of course, Kenya would have a far better chance of squeezing (still with some strain) through the bottleneck if it were to get on with its family-planning campaign with far greater energy than to date. If the country were to achieve replacement fertility (more or less the two-child family) by 2010 instead of 2040, its eventual population would be (only) 67 million people.[16]

Not surprisingly, one-quarter of sub-Saharan Africa's people today experience what is officially known as "food insecurity." That is, they do not have enough food even in normal (nondrought) years to allow them to pursue an active working life. Or, in the case of countries with an adequate food supply overall, many people do not have enough money to buy it.[17]

Fortunately there is some scope to expand food production, primarily through irrigation, of which only one-quarter has been developed.[18] The largest need for irrigation water is during the dry season, when water availability can fall as low as one-tenth of annual flow. But in a region which, as we have seen in Chapter 3, is unduly dry at the best of times, water deficits are already a pronounced problem, and likely to grow worse fast.

The Population Factor

The region's population is projected to increase from 550 million today to an ultimate total of more than 2.6 billion before zero growth is finally reached well over one century hence. (The AIDS scourge at its appalling worst will not cut population growth by more than one-third.) This is quite the largest proportionate upsurge in human numbers anticipated anywhere on Earth. Asia is expected to expand its numbers (only) 2.5 times and Latin America 2.6 times. The difference is largely because there is still a cultural premium on large families in Africa. An average woman produces six children, while eight is often the norm. In Asia and Latin America, by contrast, the average number of children is only four. Moreover, it will be difficult to do much to reduce population growth sharply for decades if only because of "demographic momentum." In the wake of explosive growth in the recent past, over half of all Kenyans, for instance, are under the age of fifteen (in North America, only one-fifth), which means a large number of potential parents are already born. If Kenya were to attain the two-child family forthwith, its population would still grow for another two generations and would eventually double before leveling off at zero growth.

Population growth of this order makes it far more difficult to feed everyone. While there has been about a one-fifth increase in total food output in the region during the 1980s, the potential gain per head has been more than wiped out by the upsurge in human numbers.

Nor can the future be considered hopeful on the population front. A minimum target of a 50 percent decline in the number of children per woman is not likely to be reached in less than three decades, and it would still do no better than bring sub-Saharan Africa into line with what Asia and Latin America have already achieved. Even under this scenario, the region's population would top 1.2 billion by the year 2020.

As too many people press on too few croplands already, the upshot is the growing incapacity of the environmental-resource base to sustain still more people—people with a growing sense of desperation. A further result is apt to be the creation of first-rate breeding grounds for disaffection, disputes, and revolts of multiple sorts. Lack of food—the most basic of all needs—eventually serves as a major source of frustration, civil disorder, and outright violence. Already there have been food riots in urban communities of Ghana, Sudan, Ethiopia, Uganda, Angola, Zambia, Mozambique, and Zimbabwe.[19]

Better News

Fortunately there is some better news. On the population front, three-quarters of Africans now live in countries with governments that view their population growth rates as too high.[20] There is still much opportunity to reduce the long-term problem. Were Nigeria, for example, to implement its new family-planning program with due vigor, it could conceivably achieve an average of two children per woman (replacement fertility) within thirty years. If this rapid fertility decline had begun in 1985, there would be only 195 million Nigerians in the year 2015, and 270 million by the time that growth finally fades away a century hence. Since the rapid fertility decline could not have begun before 1990 at best, the totals will expand to 219 million and 327 million; and if the fertility decline is delayed until 1995, then 239 million and 400 million. In other words, a delay of only ten years from 1985 to 1995 will make a difference of 130 million people in the ultimate size of Nigeria's population. In 1982 when the country still had fewer than 100 million people, Nigeria felt so overburdened that it summarily expelled between 2 million and 3 million aliens, causing large-scale human misery and generating acute tensions with neighboring countries.

On a regionwide scale, if women average two children each by 2030, the ultimate population will be 1.4 billion people. But if the two-child average is delayed until 2065, a gap of only thirty-five years, the ultimate figure will be more than three times greater, 4.4 billion.[21]

In 1990 only 3 or 4 percent of reproductive women were practicing contraception, compared with at least 30 percent in India and 70 percent in China. It costs around $20 to supply a contraceptive user for one year. So if 25 percent of at-risk women in the region are to be supplied by the year 2000, there will be an annual bill of $640 million. But thus far family-planning budgets amount to less than $100 million—and over half of that comes via foreign aid, while constituting a bare 1 percent of all aid to the region.

Fortunately, there is a success story to point the way ahead. From 1982 the Zimbabwe government has been using an array of tax incentives and other inducements to promote family planning. As a result, Zimbabwean women increased their rate of contraceptive use from 14 percent in 1982 to 38 percent in 1987, an exceptional achievement in such a short time. By 1990 the family size had been cut by two children.

On the environmental front, too, many new initiatives are under way to reduce soil erosion, desertification, and deforestation in Zimbabwe. In conjunction with policy reforms in the agricultural sector such as price incentives, reduced subsidies, better extension services, and expanded marketing networks, there was an annual increase in agricultural production of 3 percent during the late 1980s.[22] Zimbabwe has shown what is possible by turning from a food importer as recently as 1982 into a food exporter in 1987. Regrettably this success story has recently been reversed due to severe drought throughout southern Africa, revealing the fragility of apparent success under environmental pressure.

Sub-Saharan Africa
and the North

To complicate the picture, we must recognize that future progress depends in major measure on the international economic order. The outlook is not promising. In just 1986, for instance, declining prices for commodities such as copper and coffee removed $19 billion from the region's trade revenues, or about four times the amount which the region was promised in emergency aid during that drought-stricken year.[23]

Still worse is the debt problem. The region owes around $160 billion, or twenty-five times more than in 1970, and the seventeen most indebted countries face annual payments of almost $7 billion. Many African countries spend half of their export earnings on debt-interest payments, and some must divert virtually all their export earnings to this end. (In 1985, for every dollar the West donated to famine-stricken African nations, it took back two dollars in debt repayments.) These economic straits make it increasingly difficult for the region to attract development capital from advanced nations, further frustrating efforts to reduce poverty. Together with decreased financial flows from international lenders, the upshot has been a net decline in cash transfers from $10 billion in 1982 to less than $1 billion in 1991.

Fortunately there is a bright star in this dark picture. The United States, Canada, Britain, and Germany have recently forgiven outstanding debt of the most impoverished nations in the region, worth a total of more than $1 billion, against promises of economic reform. Were other donor nations to follow this lead, it would save the region between $300 million and $400 million a year—albeit out of debt-servicing needs of over $30 billion. Regrettably on the aid front the region fares dismally. The United States supplied $800 million in 1991, only one-third as much as to a single North African country, Egypt.

Refugees

I have already noted, in Chapter 4, the situation in Ethiopia. In the Ivory Coast at least one-fifth of the population are unofficial immigrants from the Sahel, causing an acute problem of deforestation due to land hunger: one-third of landless people are immigrants. In Sudan there are well over one million refugees from outside, adding to the economic burdens and political instability of that overstrained country—where almost another one million people have had to flee their homes for environmental reasons (and 8 million face starvation). In both the Ivory Coast and Sudan the primary reason for the refugee phenomenon is desertification; in the region as a whole, people displaced by drought were estimated to total at least 3 million as long ago as 1986.[24] By the year 2000 there could well be an additional 70 million people affected to some degree by desertification.

All this has important implications for security in the nations concerned. It is hardly coincidental that not one government in the main desertifying zone, the Sahel, survived the succession of droughts during the 1970s, and many fell twice—while Sudan is moving toward a third collapse in ten years.

Counting all refugees in the region, we find the total has grown from less than 1 million in 1970 to 3 million in 1980, 10 million in 1985, and 14 million in 1990, or well over half of all refugees worldwide. This recent outburst is not surprising in a region where (as noted above) in 1985 some 150 million people, or one-third of the total, faced food shortages, and where 30 million suffered from famine: a result not so much of recent drought as of long-standing environmental degradation compounded by ultra-rapid population growth and faulty development policies. Although there are no accurate breakdown figures, I believe, on the basis of twenty-four years' residence in the region, that many more refugees deserve to be classified as environmental refugees rather than political refugees.

War in Africa

As noted, civil wars and other violent conflicts have been all too frequent. At the same time, and largely as a result of increasing strife, there has been a steady "militarization" of the region. During the 1980s military spending expanded six to ten times as fast as the region's economies were growing. In certain years the value of grain imported was exceeded by the value of arms imported.

In part, the region's wars have derived from political and ethnic factors. But in part, too, they have reflected the desperation of communities impoverished beyond endurance, as in the notable cases of Ethiopia, Sudan, Chad, Angola, and Mozambique. All these rank among the most underdeveloped nations of the region. All feature fast-growing populations. All suffer from severe environmental degradation. All have experienced prolonged political instability. All are subject to periodic military upheavals. And all feature some of the highest per-capita expenditures on military activities. Every African nation experiencing civil war has also suffered from drought, made worse if not caused by environmental abuse. The numbers of people suffering food shortages have reached 7 million in Ethiopia, 8 million in Sudan, 3 million in Mozambique, and 2 million in Angola, or 20 million out of 34 million such people in the region as a whole.

It would overstate the case to assert that environmental problems, together with associated problems such as burgeoning poverty and population growth, have been the predominant or even the prime factor in each of the countries listed. But there is a strong connection. In Mozambique, perhaps the worst affected country, farms have been producing only around

70 pounds of cereal per citizen, while food aid has been totaling almost as much. At least one person in four has been affected by food shortages, one person in eight has had to migrate from one part of the country to another, and one person in fifteen has fled the country altogether.

These, then, are some major factors in the complex of problems that interlink population growth, environmental degradation, economic decline, and growing political instability, culminating in a spread of conflict and violence. The outlook is far from positive: recall the quotation at the start of this chapter. Without greatly increased relief measures both within the region and from outside, the prospect must be a continuing deterioration in the security situation, with repercussions that will hardly be confined to sub-Saharan Africa:

> The picture for the period ahead is almost a nightmare. The potential population explosion will have tremendous repercussions on the region's physical resources such as land and essential social services. As a result of socioeconomic difficulties, riots, crime, and misery will be the order of the day. With weak and fragile sociopolitical systems, the very sovereignty of African states will be at stake
>
> [Dr. Adebayo Adedegi,
> executive secretary of the
> Economic Commission for Africa, 1983].

Scenarios

A downside scenario foresees that the region continues to suffer from the dire triad of adverse environments, food shortages, and poverty, all aggravated by runaway population growth. Some outsiders say they have poured mountains of money into the region without much progress, so what's the point in more? There's got to be a limit or you start to suffer from compassion fatigue. Some, notably those who pride themselves on being "practical" in a difficult world, assert that if the region disappears into an abyss of permanent poverty, what is that to the rest of the world when the region accounts for less than 1 percent of the global economy? Critical minerals have been stockpiled in the industrialized nations, and there are no longer any strategic communication lines at stake. So let the region fade from the radar screen of concern. Still others react that if Africans want to go on making war on each other without an end in sight, why should outsiders try to stop them? A public opinion survey shows that many outsiders cannot immediately locate Ethiopia or Mozambique on a map.

In the wake of this indifference, one African country after another revokes on its debt, which hardens outsider sentiment. A common prognosis is that the region's plight will get worse before it gets much worse. Poverty, already more extreme than any known on this scale in human history, grows ever more entrenched, and living levels sink to barest survival. The new rulers of Libya, who overthrew Colonel Khaddafi because he was too soft, foment popular rebellions—here a coup, there a coup, everywhere a coup upon coup. Because the United States has allowed its oil imports to soar and OPEC has seized its chance to hike prices to new heights, the Libyan leaders use their vast oil revenues to purchase nuclear capabilities. They start to talk of striking back at the oppressive North, gaining a ready hearing throughout the region. A new security specter looms: Armageddon Africa-style.

A more positive scenario anticipates that world opinion is finally stirred by the 1995 drought disaster when 23 million Africans die and over 150 million or one-quarter of the total population slip below the starvation-risk level. Half of all debt is canceled forthwith and most of the rest is put on hold. The revelation that a number of national leaders have made themselves richer than the poorest one-fifth in their countries drives African governments to a purge on corruption. But these measures are too late. Multitudes of famished people from West, East, and Central Africa flee into the least densely populated countries, Zaire, Congo, and Gabon, where they burn entire tracts of the equatorial forest to make way for croplands. This causes the carbon emissions from all tropical forest burning to jump within just two years—a period when the grain baskets of North America and Eurasia suffer their most pronounced droughts as a result of the global warming under way.

The rich nations decide it is in their own interest, too, to stem the damage. A stirring speech by the American president at the new World Emergency Assembly declares that the overfed nations cannot sleep at night as long as the underfed endure an unprecedented tragedy in Africa. The new nation of Azania, formerly South Africa, supplies a flood of investment, technology, and development expertise to the region, and a dozen countries as diverse as Zimbabwe, Kenya, and Ivory Coast achieve "economic takeoff." A Save Africa Plan mobilizes funds for a blitz program for agriculture, against an undertaking by African governments to introduce a crash campaign to slow population growth. The African Development Bank announces that the region should be able to get back to 1960 levels of economic wellbeing and nutrition by 2010. Plans are initiated for foreign capital to exploit the immense mineral riches of the region, igniting hopes that within another generation the region will finally play its proper place in the global community.

6

The Philippines

> Deforestation and other forms of environmental degra-
> dation are one of the principal sources of social injustice
> in the country, and one of the causes of armed insur-
> gency.
>
> —Philippines Government
> Policy Advisory Group, 1987

I have visited the Philippines
four times. On each occasion I have been struck by the extraordinarily agreeable demeanor of the people, some of the most
pleasant you could wish to meet. I have also been struck by the
extraordinarily destitute state of their lives, stemming in major
measure from grand-scale rundown of their environments. In
particular, I have come across entire communities of subsistence farmers living on the limits of survival. In addition, I

recall a weekend during my last visit in 1988 when extensive sectors of the capital city, Manila, were cordoned off following an attack by guerrillas from the New People's Army on a military installation. I could not help but be impressed by the fact that this small but fiercely committed group could penetrate with its assault to the very heart of the government. It was partly explained by the equally impressive fact that many citizens were outspoken in their support for the guerrillas on the grounds that they felt the government was largely insensitive to their impoverishment.

This was all the more significant in that the Philippines was then playing a salient role in geopolitics. Two very large bases stood at the center of the United States' military operations in the Pacific Ocean, serving as a springboard for the projection of American power throughout much of Asia. It had become still more important since the mid-1970s changes in Vietnam, in particular protecting the three Indonesian straits through which was shipped much Middle East oil. True, there were alternative sites for bases elsewhere in the western Pacific, and these are now being taken up by the United States following its departure from the Philippines. But this option is more expensive and less effective militarily.

In any case, the United States in common with other world leaders wants the Philippines to become a prominent example of democracy and economic security in the proper broad sense. Regrettably the Philippines is anything but secure. There have been a dozen coup attempts in the past few years, all aimed at overthrowing the America-allied government. And the Philippines will continue to be a country in deep disarray as long as its economy remains shaky and political stability elusive. Unless it can enjoy an extended phase of sustainable development, with economic advancement especially for the peasantry, there will be growing support for the New People's Army and other dissident groups that would like to see an end to Western-style governments.[1]

The economic outlook is unpromising, largely for environmental reasons.[2] A sizable share of the economy is environmentally based. The sectors of agriculture, forestry, and fisheries contribute a large proportion of GNP and export revenues, and employ over half the labor force. The bulk of the population lives in the countryside, made up of subsistence peasants on the fringes of the main economy; a full 70 percent of Filipinos depend on agriculture and fishing for their livelihoods, and almost as large a share suffer from one form or another of chronic malnutrition. Yet the natural-resource base has been widely depleted through deforestation, soil erosion, watershed abuse, disruption of water systems, overharvesting of fisheries, and destruction of coral reefs. Foremost among these is deforestation, which is the initial cause of several of the other problems. So this chapter focuses on deforestation in the Philippines, before going on to look at how deforestation fosters insecurity in a number of other countries.

There are two central factors of the Philippines' environment, and of its economy as well. First is the shortage of farmlands. Nationwide the amount of arable land per rural inhabitant has declined to less than one acre, which means that an average family of six persons must try to sustain itself off little more than five acres. Even these small amounts may well fall by the year 2000 to only 0.6 of an acre per head and 3.5 acres per family. Yet to cope with population growth and nutritional needs, the country must achieve a 4 percent annual increase in agricultural output. It achieved better than that from 1965 to 1980, but for the past decade the rate has slumped to a mere 2 percent. Worse, the 4 percent goal has now become all the more a questionable goal given recent environmental trends.

The second factor, closely related to the first, is that more than half the country is hilly or mountainous. This, together with friable volcanic soils and heavy tropical downpours, makes farmlands unusually vulnerable to soil erosion. Almost half the uplands, moreover, feature slopes so steep that they are quite unsuitable for agriculture without stringent soil-safeguard measures. Remaining uplands areas for agricultural settlement are fast being claimed by a rising tide of immigrants from the overcrowded lowlands. While the "agricultural frontier" (meaning the cultivable lands available) was closed a good while ago in the lowlands, a similar frontier will shortly become closed in the uplands too—bringing on the prospect of a rapid increase in cultivation of steep slopes. In other words, there is a "threshold effect" impending. While population growth has ostensibly been sustainable to date, albeit with certain long-term costs, it will shortly induce exceptionally severe burdens on the environmental foundations of the national economy. Already almost half the uplands are so badly eroded that they can no longer grow much in the way of crops.[3]

Forests

These two key factors of the Philippines' environment are exemplified by the country's forests. While the Philippines was once a sea of forest from end to end, almost all remaining forests are now confined to the uplands, covering a mere one-fifth of national territory and with the most productive old-growth forests now reduced to relict patches.[4] At recent rates of cutting, these forest fragments will shortly be unable to meet domestic demands, let alone export markets. Whereas timber exports in 1967 amounted to one-third of all exports, today they are worth only a small part as much, and the Philippines seems set on becoming a timber importer within a few years—a debacle indeed, both economically and environmentally, after the country once led the world with its timber exports. In fact, there is a downright timber famine in the making. Domestic demand for timber, even if held at the current inadequate level, is projected to exceed the 1985 amount by two-thirds in the year 2000 and sixfold in the year 2025.[5]

As mentioned, the uplands are having to accommodate growing throngs of migrants from the lowlands. The uplands population now totals more than 20 million people, or almost one-third of all Filipinos, with a rate of increase half again as large as that of the national population.[6] This flood of migrants has led to much deforestation, among other forms of environmental degradation, and already a good number of farms are located on extremely steep slopes. Many of the migrants are below the official poverty line, so they cannot afford conservation technologies and other farming investments that offer only long-term payoffs.

Population projections indicate that by the end of the century, when the present Philippines populace of 64 million people will have expanded to 74 million (and more than 100 million by 2025), all available arable land in the uplands will be occupied by small-scale cultivators, and a large amount of steeply sloping land will be under cultivation as well. In turn, this means that virtually all remaining forests will have been eliminated. In turn again, this means that the country's major hydrological systems will have been seriously impaired, with ecological repercussions such as flooding and river sedimentation reaching across much of the lowlands.[7] Most of the 1991 flooding disaster that left 2000 dead (some observers say three times as many) and half a million homeless was attributed to uplands deforestation, leading to huge mudslides downstream. All in all, the mass movement of people into the uplands, extending over just a few decades, may eventually come to rank as a pivotal event in the Philippines' entire history.

Deforestation is especially adverse for those upland watersheds that are critical for water supplies to feed hydropower facilities, irrigation projects, and domestic needs in urban areas.[8] Sedimentation of outsize dams, due to soil erosion in the wake of uplands deforestation, is cutting short the operational lives of hydropower reservoirs by half or more. River flows are being reduced and becoming irregular, which is all the more adverse in light of the country's plans to expand irrigated agriculture.[9] During the last quarter of this century, demand for irrigation water is projected to double if the country is to expand its agriculture enough to keep up with growing numbers of people and their growing appetites. Output of the most widely irrigated crop and the country's leading staple, rice, no longer keeps pace with population growth. Although the country became self-sufficient in rice by 1974, it has had to rely on imports again since 1985. So there is a premium on efforts to safeguard water catchments through reforestation. But fewer than 80 square miles are reforested each year, out of at least 400 square miles required.

Other Environmental Problems

The country faces water shortages for agriculture in still broader senses. During the past quarter century the areas affected have expanded 17-fold.[10] In 1983 the country experienced a modest failure of the monsoon, whereupon it suffered the most extensive and protracted drought effect in thirty years, with 5600 square miles of agricultural lands affected and a loss of one-tenth of the usual harvest.

Next, fisheries. With a coastline longer than that of the United States, the Philippines possesses plenty of offshore waters with abundant fisheries.[11] The annual catch, now around 2 million tons, generates greater fish food per head and a greater proportion of animal protein than is the case for any other Southeast Asian nation. The present catch is reckoned to be the maximum that can be taken without overexploitation of fish stocks. Yet by the year 2000 the projected total of 74 million Filipinos is likely to be demanding more than 3 million tons of fish, and by 2025 more than 100 million people will likely be demanding over 4 million tons. Already there is much overfishing. Certain stocks have been reduced to less than one-third of what they once were, with an economic loss reckoned at more than $90 million annually.[12]

Population

All of these problems are aggravated by pressures from a population that has more than tripled since 1950. The Philippines is the only Asian country where the birth rate has been rising for much of the recent past. Because of church pressure in this Catholic country, the government has been little inclined to do much to reduce the growth rate. Fortunately there are now signs that certain political leaders have accepted that the country's economic recovery is being stalled by population growth, among other factors, and the government plans to tackle the problem through, for example, a $25 million United Nations program to support family-planning services.

But it is an uphill battle. There is reduced motivation for parents to practice family planning when 150,000 children under the age of five die each year, largely from diseases that reflect malnutrition and dirty water. In turn, it is hard for the government to assign additional funds to health as long as the country is saddled with a debt burden, $30.5 billion, that is equivalent to well over twice all annual export revenues. It is within this context that, as the United Nations Children's Fund points out,[13] foreign debt ceases to be a matter for a few financial experts and enters the daily arena of life and death for tens of millions of people.

At the same time, let us bear in mind the other factors at work in the Philippines' faltering economy in addition to environmental decline and population growth. Included is the inequitable division of farmlands. Just 3 percent of landowners control one-quarter of the country, while 60 percent of rural families either cannot survive on their tiny plots or have no land at all.[14] Then there are faulty development policies, bureaucratic inertia, and corruption; during the last ten years of the Marcos regime, the president and his wife were salting away funds equivalent to all the annual economic advancement of the country as a whole. But environmental and population problems make other problems worse, and seem set to make them much worse in the future. If there remains little prospect of economic advancement for the majority of Filipinos (real per-capita income fell by 20 percent during the 1980s), growing throngs of disaffected peasantry will increase their support for rebel groups such as the New People's Army.[15]

Anti-Government Rebels

Insurgents constitute a growing threat to the government, speeding the tempo of unrest all around. After twenty years of building up their strength, they now operate in most provinces and they exert outright control over one-fifth of the entire country. Periodically they penetrate into downtown areas of all major cities. They have been mounting hundreds of attacks against the government each year, despite intensified counteroperations.[16] In particular, the government's "hearts and minds" campaign has made little headway with rural communities, disenchanted as they are with the degradation of the environmental basis of their livelihood. Rather, they provide support for the insurgents, thus enabling 25,000 rebels to gain ground against ten times as many government troops.[17] To this extent, environmental rundown, especially deforestation, serves to foster political instability in the Philippines.

Deforestation
and Security Elsewhere

In other parts of the tropics, too, deforestation leads to economic dislocations and political instability, if not confrontation and conflict. In India, local communities, witnessing their forest homes targeted for official felling, have frequently taken on the government, albeit through passive resistance for the most part. In the Sarawak sector of eastern Malaysia, forest tribes who also watch their homelands disappearing before government-sponsored chainsaws regularly engage in head-to-head opposition to officialdom. More serious still, disaffected forest peoples provide support for rebel forces in northern sectors of Thailand and Myanmar (Burma). Similarly, in Peru, where three-quarters of the rural population is landless or possesses too little land to sustain a farming livelihood, the forests and the impoverished peasantry of Amazonia supply a base for the Shining Path insurgents. There are other instances in Indonesia, Colombia, and Guatemala, whether large or small in scope, and we can expect them to multiply as forestlands feature fast-growing multitudes of destitute peasants.

As we have seen through the Philippines' experience, deforestation causes poverty, just as poverty causes deforestation. On a larger scale, and as we will see in Chapters 7 and 8 dealing with the Indian subcontinent and El Salvador, deforestation leads to the sedimentation and silting of reservoirs, irrigation channels, natural lakes, harbors, and offshore waters. In turn, these repercussions adversely impinge upon hydropower installations, irrigation agriculture, domestic water supplies, port facilities, and fisheries both inland and coastal. All this acts to undercut economic prospects and to accentuate poverty, with all the destabilizing consequences that engenders.

This is especially the case as concerns the clogging of hydropower facilities. These energy sources will, it is hoped, account by the late 1990s for a full two-fifths of developing countries' electricity output, at an investment cost of well over $10 billion. But an annual sedimentation rate of 2 percent, as is revealed by 200 major dams built from 1940 onwards, means that the operational capacity of these dams will be reduced by one-third by the year 2000. Just a 1 percent capacity reduction would mean a loss of almost 150 gigawatt-hours of electricity. To produce an equivalent amount of electricity by thermal means would require 37 million tons of oil. At $160 per ton, or $25 per barrel, the sedimentation-derived loss would result in the loss of $6 billion of electricity output in the single year of 2000.[18]

Still further consequences arise from deforestation, this time through linkages that are less direct and more diffuse. When forests are depleted through excessive gathering of fuelwood, among other forms of deforestation, rural households divert animal manure and crop residues from farm fields to house hearths. So fertilizer benefits have to give way to fuel needs, even though cropland fertility declines. In the developing countries of Asia and Africa alone, at least 400 million tons of dung were burned each year during the early 1980s. Each ton meant a loss of at least 110 pounds of potential grain grown, or a total of 20 million tons of grain per year (some analysts calculated the total at twice as much).[19] To put this calculation in perspective, 20 million tons of grain can feed 100 million people for one year, or they can make up the likely shortfall in grain requirements of 500 million people. To purchase the grain on world markets would cost $3.5 billion—a figure to bear in mind when we consider the cost of establishing fuelwood plantations, estimated at roughly $1 billion a year for ten years.[20] In this way too, deforestation serves to undermine economic prospects and to intensify poverty.

Deforestation can even cause problems in another country altogether. As we have seen, decline of forest cover in upland catchments sometimes leads to so much disruption of watershed systems that it generates flash floods downstream—occasionally a long way downstream. Deforestation in the Ethiopian highlands contributed to the extraordinary events in central Sudan in mid-1990, when Khartoum, at the confluence of the two main feeder rivers of the Nile, received the equivalent of two years of rainfall in a single day. This freak phenomenon occurred at the time when the Blue Nile from Ethiopia was more than full already. The result was flooding that caused massive economic damage.

Yet another deforestation linkage arises, this one still more important in a general sense. Well over half of all deforestation is now due to the displaced peasant or the landless farmer who finds himself impelled to migrate from traditional farming territories into the last unoccupied lands available, forests.[21] There he takes up a slash-and-burn life-style, felling vast tracts of forest. As long as there were not many of these forestland farmers and there was plenty of forest available for them to roam around in, they could practice a migratory form of small-scale cultivation, allowing the forest plenty of time to restore itself. It made for a self-renewing form of forestland farming, sustainable over centuries.

How different is the situation today, when many more farmers are practicing slash-and-burn cultivation in only half as much forest. The shifting cultivator has given way to the "shifted cultivator," who destroys the forest outright. This pattern has been noted in the Philippines, and it is strongly established, too, in Bolivia, Peru, Ecuador, Ivory Coast, Madagascar, Myanmar, Thailand, and a dozen other countries.[22] It will become still more widespread if only because of population growth. By the year 2030 or thereabouts, there will be another 3 billion people on Earth (or as many as all humankind just three decades ago)—and the bulk of them will have been born in tropical forest countries.

The shifted cultivator now accounts for more deforestation

than the commercial logger, the cattle rancher, and all other agents of deforestation combined. So fast are their numbers growing that these displaced peasants may well be accounting for three-quarters of all deforestation by the end of this decade or shortly thereafter.

Clearly their role is crucial not only to the forests' survival but to economic prospects and hence to political stability. It is worthwhile, then, to take a closer look at their situation and how it has arisen. It turns out that they are driven into the forests by a host of pressures, the predominant one being land hunger. While often reflecting population growth, this factor can stem as well from the unfair division of traditional farmlands. In Brazil, 5 percent of farmers occupy at least 70 percent of arable lands, some with holdings so large that they cultivate only half, while the hardscrabble peasant over the fence has to make do with just a few acres. Largely as a result of this inequitable division of farmlands, 12 million rural Brazilians are landless or do not have enough land to survive on. At the same time, 108,000 square miles of farmlands lie idle, enough to accommodate 7 million ten-acre smallholdings—or more than all small-scale farmers in the whole of Brazilian Amazonia.[23]

So a key to stemming the flood of shifted cultivators lies with an esoteric-sounding measure, agrarian reform, referring to a fairer division of traditional farmlands. Also helpful would be more government support for subsistence agriculture (by contrast with expensive Green Revolution agriculture) in order to help the peasant make more intensive use of his diminutive plot back home. In addition, the small-scale cultivator stands in acute need of further forms of assistance such as better credit facilities, extension services, marketing networks, and a host of other rural-development measures that would relieve his motivation to pick up his machete and matchbox and head for the forests. To this extent, the shifted cultivator represents a failure of development strategies across the board, and his problem can be confronted only by a major reordering of broad-scope policies on the part of governments concerned. Something similar applies to aid agencies. British prime minister Margaret Thatcher allocated £100 million to help Brazilian Amazonia, but it almost all went to improved logging, more parks, greater forest research, and other initiatives that addressed symptoms rather than sources of the essential problem.

Still other factors are involved. Tropical forest countries would feel more inclined to spend more on subsistence agriculture if they were not saddled with foreign debt. Brazil owes $116 billion, or around $760 per citizen. During the mid-1980s, Brazil was making debt repayments that in relation to GNP were twice as large as those of Weimar Germany. To this extent, some of the most efficient deforesters are on Wall Street, in the City of London, and among the gnomes of Zurich. Yet who would bet that many of them give a thought to the added effects of their financial calculations across the oceans?

It may sound strange to urge the cause of land redistribution and similar non-forestry measures in territories many horizons away from the forests, and debt relief in financial citadels overseas, as key strategies for safeguarding the forests. But the forests are largely being destroyed for non-forestry reasons— nothing to do with the need for better logging practices and other traditional responses. This helps to explain why there is all too little attention directed to the root causes of deforestation by the major save-the-forests initiative to date, the Tropical Forestry Action Plan. It is essentially a forestry strategy, by contrast with the measures required—population planning, agrarian reform, agricultural advancement—to address broader-scope issues of development (or, rather, maldevelopment).

In fact, it is an indication of how far the shifted-cultivator problem has been systematically neglected; while it has been recognized in principle for ten years, it has been almost entirely disregarded in practice. We have no worthwhile idea of how numerous they are; we have no better assessment of the total than somewhere between 250 million and 500 million people. If the latter figure is correct (it may well be an underestimate), the shifted cultivator amounts to almost one in ten of humankind. Yet he remains a forgotten figure. Nor should he be viewed as some sort of "culprit." He is no more to be blamed for torching the forest than a soldier is to be blamed for fighting a war.

This all means that unless we do a far better job of tackling their plight in "back-home" areas of traditional farming territories, there will be a fast-growing reservoir of impoverished peasants in deforested lands. They could well total one billion people shortly after the year 2000. Disillusioned as they will be at their governments' ostensible indifference, they will supply fertile breeding grounds for widespread discontent and anti-government resentment, plus support for insurrection and revolt. Thus they represent a powerful source of potential trouble. Yet they remain ignored for the most part. Few political leaders recognize their linkages to national economies, let alone their likely contribution to future conflict.

Scenarios

A downside scenario foresees that as environmental decline worsens, it is paralleled by political decay and economic disaster. The last upland forests give way to the commercial logger and the shifted cultivator, whereupon soil erosion increases faster than ever. This depletes the operational capacity of several large reservoirs, aggravating the country's energy problems. It also reduces the efficiency of several irrigation networks that are vital to lowland agriculture. Food production shows no increase year by hungry year. The lowlands can support fewer rather than more people than before, and migration into the uplands increases yet again, fueled by uncontrolled population growth. The overall upshot is economic calamity:

zero growth for GNP, leaving the impoverished throngs to
grow ever bigger—and more desperate.

In the meantime the government persists with the elitism
and special-interest chicanery that have characterized the past
three decades. Corruption returns on a scale to match the Marcos years. The majority of the populace feels ignored and frustrated: no signs of hope that their plight will be recognized, let
alone addressed. The ranks of the New People's Army swell
and the enlarged insurgency enjoys broadening support in
countryside and cities alike. The NPA becomes more daring as
government forces grow demoralized. In 1997 it launches a
nationwide campaign to topple the government. Fortunately
for the government, outsiders intervene to help "stabilize" the
situation—which leads to still wider and deeper-rooted instability. The contest remains inconclusive for a while, with escalating costs to both sides. Eventually the government forces
are so thinned through defections to the enemy that the end
arrives swiftly. After a bloodbath week of "settling scores by
the score," a new populist government emerges. But the devastation to both the environment and the economy has been so
extensive that most of the country slips deeper into poverty
without prospect of respite.

A positive scenario foresees that the government finds the
rebels becoming stronger by the month. It realizes there must
be a negotiated resolution. The 1994 peace conference, sponsored by the United States and Indonesia, brings together the
two sides in an "anything goes" gathering. A new government
is set up with democratic rights assured for all communities of
whatever sort. This has been a key demand of the 2500 citizens
groups and other non-governmental bodies that have sprung
up in recent years as people despaired that the government
would never look beyond its elitist and corrupt cronies.

The new leaders inspire a mass mobilization of all Filipinos
to rebuild the country they now feel has become their country
once again. They direct priority attention to reforesting the

uplands and restoring watersheds, while rehabilitating the many other environments that have been loaded beyond carrying capacity. A nationwide population-planning effort soon brings down fertility to match many other Asian countries, and the accepted aim is to bring forward the date of "replacement fertility" from 2035 to 2010 at the latest—a goal that would shift the country from the most "backward" in Asia as concerns population planning to rank among the leaders.

By 1996 the Philippines needs to import only a few essential foods. Plantations of fast-growing trees will be ready for harvest by shortly after the year 2000, and there is the prospect that the country could eventually reclaim its place as Asia's number one source of specialist hardwoods and generate a full one-third of export earnings. Economic growth surges as the environmental base is regenerated. Political stability follows close behind. The entire country, assured at last of its all-round security, rallies to the cry of "Filipinos first!"

7

The Indian Subcontinent

Water supplies available for irrigation are already short of optimum crop requirements. Even if all rainfall supplies and groundwater reservoirs are exploited, water will still fall far short of future requirements.

—Ahmed Mohtadullah,
Director, Pakistan Water and
Power Development Authority

Deforestation has brought our nation face to face with a major ecological and socioeconomic crisis.

—Former Indian Prime Minister
Rajiv Gandhi

By 2010 Bangladesh will be heading towards a major environmental crisis.

—Dr. Fasih U. Mahtab,
Chairman of Planning and
Development Services, Bangladesh

This region (Figure 7.1) presents much scope for conflict. Many root causes lie with environmental factors in the form of deforestation, soil erosion, depletion of water supplies, and desertification. This environmental decline is leading to agricultural setbacks, indeed to the

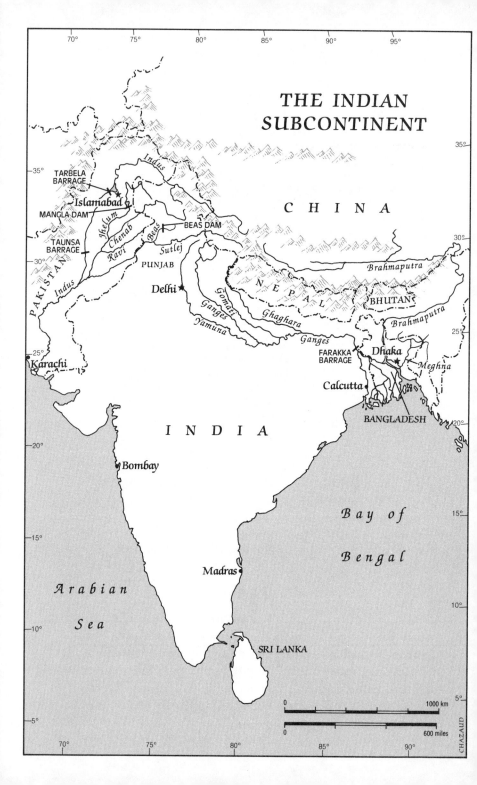

growing incapacity of many areas to support human communities. Yet the total population of the region, more than 1.2 billion people today, is projected to approach 1.3 billion by the year 2000, 1.9 billion by 2025, and 2.6 billion before it finally levels out at zero growth late in the next century (Table 7.1). In the meantime the region is expected to experience pronounced climatic dislocations through global warming with its sea-level rise and its disruptive impact on monsoon rainfall.

Important as this is within the subcontinental context, it is yet more significant in light of the region's strategic location in the broader international arena.[1] The subcontinent has historically ranked as a zone of marked interest to other Asiatic nations, certain of which, notably Iran and China, believe they might benefit from a period of increased instability in the region. In addition, the subcontinent dominates the Indian Ocean, which, after remaining largely unmilitarized to date, has been experiencing some naval buildup.

Moreover, the subcontinent features two religious communities, Hindu and Muslim, which have long been antagonistic toward each other. This religious confrontation, a source of persistent tension and periodic hostilities among Pakistan, India, and Bangladesh, may shortly be exacerbated by the fundamentalist Muslim spirit that emanates from Iran and elicits sympathy among certain Pakistani factions. There may be further religious linkages to Muslim-dominated states in central Asia (the former Soviet Union), some of which are showing more volatility than ever.

In this chapter we shall concentrate on Pakistan, India, and Bangladesh, since these countries comprise roughly 95 percent of the region's area and its population (other countries are Sri Lanka, Nepal, and Bhutan). In each country the natural-resource base is central to the economy, in which agriculture is the largest sector. Pakistan's agriculture employs well over half the populace and earns four-fifths of export revenues; in India, four-fifths and two-thirds; and in Bangladesh, three-quarters and four-fifths.[2]

TABLE 7.1 INDIAN SUBCONTINENT: SELECTED BASIC INDICATORS

Country	Area (sq. miles)	Population Mid-1992 (millions)	Population Growth Rate 1992 (%)	Projected Population (millions) Year 2000	Projected Population (millions) Year 2025	Projected Final Population (millions)	Per-Capita GNP 1990, U.S. $ (and average annual growth rate 1965–90, %)	Average Annual Growth of Agricultural Output 1980–90 (%)	Foreign Debt as % of GNP 1990
Pakistan	310,400	122	3.1	147	281	399	380 (2.5)	4.3	52.1
India	1,269,340	883	2.0	1006	1383	1862	350 (1.9)	3.1	25.0
Bangladesh	55,600	111	2.4	128	212	257	200 (0.7)	2.6	53.8
Totals	1,635,340	1116		1281	1876	2518			

Sources: Population Reference Bureau, World Population Data Sheet 1992 (Washington, D.C.: Population Reference Bureau, 1992); World Bank, World Development Report 1992 (Washington, D.C.: World Bank, 1992).

All three countries are singularly dependent on water catchments, hence on the forests and soil cover of the Himalayan foothills.[3] But whereas earlier in this century these forests protected at least three-fifths of the foothills, ten years ago they had been reduced to only one-quarter as much.[4] As a result of the ensuing soil erosion and sedimentation of water courses, the riverbed of the Ganges River is rising by two to three inches a year, and that of the Brahmaputra by almost four inches. In the Brahmaputra watershed, the current rate of soil erosion from the Himalayas is estimated to be five times as much as in the geologic past.[5] As we shall see, this sediment loading is a prime factor in the fast-growing phenomenon of flooding.

Let us now consider each of the three countries in turn.

Pakistan

Of Pakistan's 310,400 square miles, one-fifth are irrigated for agriculture. Making up three-quarters of all agricultural lands, they produce four-fifths of the country's food. It is this irrigation network, the largest in the world, that has enabled Pakistan to grow pretty much all the wheat and rice it needs. Thus water supplies for irrigation, and for electricity and industry, are a mainstay of the national economy.

But irrigation demand is already bumping its head against the water-supplies ceiling. The water derives from the Indus River and its tributaries the Jhelum and Chenab rivers. Diversion of riverwater has steadily increased until it now comprises two-thirds of these rivers' annual flow.[6] It can hardly be expanded further without sizable seawater intrusions into the Indus River's lower reaches.[7] There is limited opportunity for greater use of irrigation water, since recycled flows are no more than half the initial withdrawals due to grossly declining quality of the water as irrigation channels pick up salts, pesticides, and toxic materials from croplands. Three times during the 1980s Pakistan suffered severe shortages of irrigation water.[8]

Equally to the point, the watershed catchments are being degraded through deforestation.[9] This leads not only to a decline in irrigation water but to soil erosion, sedimentation, and wet-season flooding. In the first half of the 1980s, floods caused losses of crops, buildings, roads, and other infrastructure worth $3 billion.[10] Because of sedimentation, the Mangla and Tarbela reservoirs, two of the largest in the world, are expected to remain operational for only a fraction of their anticipated lifetimes.[11] The reservoirs' depletion will cause water and electricity deficits so extensive that they will inflict major damage on both the country's agriculture and the economy as a whole.[12] Yet despite two decades of watershed-management efforts, the sedimentation rate continues to climb. According to the director-general of the Pakistan Forestry Institute, and referring to deforested watersheds, "The country already faces a grave situation as concerns water for irrigation and power generation."[13] Yet as a measure of growing demand in just the energy sector, the nation's need for electricity has been expanding by an average of 12 percent per year since the mid-1970s, meaning it doubles every six years.[14]

The main watershed problem lies with the decline of tree cover and its "sponge" effect. Virtually all the country's forests are in the Himalayan foothills, that is, in the Indus catchment. But only 3 percent at most of national land is forested, even though the government calculates the amount should be at least 15 percent.[15] As a result of deforestation, the country imports almost one-third of its timber, an amount that is projected to double shortly into the next century. Much more important than commercial logging as a source of deforestation is fuelwood gathering. Fuelwood accounts for half of the country's energy requirements and nine-tenths of wood consumption, and demand is expected to double within another fifteen years.[16] Yet there was an acute shortage of fuelwood as far back as 1980, when the annual growth of trees was supplying only two-thirds of the fuelwood harvest.[17]

What does all this portend for Pakistan's development prospects? The nation accomplished the remarkable feat of growing all the grain it needed by 1982, and since then it has been a grain exporter. But can the country maintain this accomplishment, given the decline of the agricultural-resource base? Agriculture faces not only uncertain water supplies, but almost three-quarters of irrigated lands suffer from waterlogging and salinization.[18]

The most important factor of all is population increase. If the present 3.1 percent growth rate continues, the amount of cultivated land per rural inhabitant will decrease from 0.8 acre in 1983 to less than half as much in 2010.[19] For two decades there has been little decline in the population growth rate of this Muslim country. A leading obstacle is that only one woman in thirty is educated. In addition, good-quality drinking water is available to only one-fifth of the rural populace—a main reason why infant mortality is higher than in adjoining states of India. As long as infant mortality remains high, there is all the less reason for couples to be motivated in favor of family planning. At the same time, continued high fertility means that demand for wheat is projected to expand from 20 million tons today to 25 million tons by the year 2000 and almost 40 million tons by 2010. In turn, this means that Pakistan will have to switch to being a wheat importer, requiring 2 million tons and 5 million tons from outside by 2000 and 2010, respectively.[20] In the latter year, wheat imports are expected to cost one-quarter more than all the nation's anticipated export revenues.

So Pakistan faces severe and growing problems with respect to its environmental foundations. Already environmental decay is leading to political strains and stresses. Year after year, acrimonious debate breaks out among the provinces over sharing of water supplies.[21] Worse still, these internal conflicts may well spill over into India. Of the Indus River basin's 400,000 square miles, 160,000 square miles lie across the border in India—a nation with which Pakistan maintains relations of armed confrontation. Regular skirmishing breaks out along much of the common frontier, especially in Kashmir with the headwaters of several tributaries of the Indus—a river from which India, too, looks likely to seek additional withdrawals of water, as we shall now see.

India

Recall that a good part of India's economy depends on the environmental base. Of national territory of 1.3 million square miles, 650,000 square miles are cultivated, and of these almost one-quarter are irrigated. It is primarily the irrigated lands that have underpinned India's success in turning itself from a severely food-deficient nation into a nation with a slight food surplus; outside the irrigated areas there has been virtually no Green Revolution.[22] The remaining farmlands are rainfed, featuring ultra-low levels of productivity—and it is in these lands that most of India's impoverished people live. By the year 2000 India will need to increase its grain production by two-fifths, yet for years only the irrigated croplands have shown any increased productivity.[23]

As in the case of Pakistan, India's irrigation-water supplies are critically dependent on forested catchments, primarily in the Himalayan foothills. Yet while the government considers that forests should cover one-quarter of national territory, forests have declined to only half as much.[24] Deforestation has led to much disruption of watershed systems, hence to downstream flooding. In the Ganges Plains alone, flooding annually has been imposing damage to crops, buildings, and public utilities averaging more than $1 billion a year.[25]

On top of all this is population growth. Today's total of 900 million people is projected to top 1 billion before the year 2000, and to surge to almost 1.4 billion by 2025 before finally leveling out at 1.9 billion toward the end of the next century. The 2025 total will mean that India will have almost caught up with China as the world's most populous nation. Whereas China has reduced its annual growth rate to 1.3 percent, India's remains at 2.0 percent, and there has been little progress in reducing it for the last ten years. True, the government spends $500 million each year on family-planning efforts, one-fifth of all such outlays in the developing world. Yet the government is reluctant to restore stronger birth-control activities after the debacle of the "overvigorous" measures attempted under Indira Gandhi's regime.

Population growth means not only extra mouths to feed. It exacerbates the problem of landlessness.[26] The total of landless households was 25 million in 1980, and it is projected to reach 44 million by the year 2000. Moreover, the total work force—four-fifths of it employed in agriculture—is expected to expand by two-thirds during the last two decades of this century, and unemployment and underemployment are not likely to improve from their present level of one-third of the work force.[27] So there is the prospect of a fast-growing pool of rural people without land or properly remunerative work. It is precisely these people—without adequate livelihoods and increasingly without hope of any improvement—who could constitute a reservoir of frustration and disaffection, ultimately leading to disorder and violence.

To reiterate a basic point, these population-and-food issues place a premium on India's people having enough water for agriculture at appropriate times of the year. While there is often too much water during the monsoon season, there is often too little during the subsequent dry season. Several parts of the country could face severe seasonal water shortages by the turn of the century when irrigation demand is projected to be half again as high as it is today and industry's demand to be at least twice as high. Already there have been clashes over water in Punjab State, where Sikh nationalists claim that too much of "their" water is being diverted to the Hindu states of Harayan and Rajasthan. Water shortages contributed to the government storming of the Sikh Golden Temple in Amritsar in 1984. Moreover, Punjab lies next to the border with Pakistan, where, as we have seen, there are growing tensions over the division of water flows from the Indus River system.[28]

Similar water-sharing conflicts have erupted over the waters of the Ganges River.[29] Rising in Nepal, where deforestation is severe and widespread, the river runs for 1700 miles before reaching the Bay of Bengal in the Bangladesh delta. Within its valleylands live more than 500 million Indian and Bangladeshi farmers. Bangladesh, occupying the lower reaches of the river, makes ever greater demands on its water flows (see below)— and the present crunch situation arises in a river basin that is projected to feature almost 200 million more people within little more than another decade.

Much of India's deforestation is due not only to agricultural expansion but to excessive fuelwood gathering. As far back as 1982, fuelwood demand was 3.5 times greater than the forests could sustainably supply.[30] The gap was met either by overcutting (thus reducing future forest growth) or by burning cow dung and crop residues (thus reducing future soil fertility). By the end of the century, the gap will be much larger as India's population continues to grow and its forests continue to contract. Even if all planned tree plantations are established, there will be a shortfall twice as large as that of 1982.[31] Already the capital city, Delhi, has to truck in much of its fuelwood from Assam, 600 miles away.

Deforestation leads to much disruption of watersheds, hence to downstream flooding.[32] The amount of flood-prone lands nationwide in 1984 totaled 230,000 square miles (one-sixth of the country), or two and a half times as much as in 1970.[33] In the Ganges Plains alone, flooding annually affects 30,000 square miles, including 14,000 square miles of croplands.[34] The value of crops lost there averages more than $250 million a year, while damage to buildings and public utilities averages around $750 million a year.[35] In addition, there is a problem of land degradation. As much as 700,000 square miles suffer from some degree of soil erosion, salinization, and waterlogging, among other adverse factors, with a marked decline in soil fertility and crop productivity in at least 400,000 square miles.[36] Soil erosion alone causes Indian farmlands to lose plant nutrients such as natural nitrogen, phosphorus, and potash that, were they to be replaced with chemical fertilizers, would cost $6 billion a year.[37] Land degradation of all kinds is estimated to cause an agricultural loss of between 30 million and 50 million tons of produce per year.[38]

Then there is a growing problem of siltation. In twenty-eight of the biggest irrigation projects with a total catchment area of 266,000 square miles, almost one-third are critically eroded.[39] In 1980, many of the reservoirs supplying these irrigation projects were filling with sediment at between two and twenty times the initial rates.[40] If only one-fifth of the live-storage capacities of major and medium-sized reservoirs are silted up by the end of the century, this will mean a loss of irrigation potential in one-tenth of the present irrigated expanse.[41] While irrigated lands were meeting the needs of only little over half of India's population in 1975, it is planned they will have to do as much for almost nine-tenths of the projected population by early in the next century.[42]

All this is not to discount India's remarkable feat in almost tripling its grain production during a period of three decades up till 1985—after which there has been all too little increase.[43] While the nation has performed exceptionally well in keeping food-output growth just ahead of population growth,[44] in the year 2000 it will need to grow half again as much grain as in the record year of 1985.[45] These growing needs notwithstanding, since 1985 there has been only a marginal increase in food-grain output. While there has been a slight increase in the productivity of irrigated croplands, many other food-growing areas have revealed an actual decline.[46]

So severe are problems of farmland degradation that by 1996 India's grain production could well plateau, with per-capita output at around 450 pounds, only marginally better than the 430 pounds of 1985.[47] But there remains much scope to intensify agricultural production.[48] India has about 20 percent more land under paddy than China, yet produces 40 percent less. As we have seen, however, the agricultural-resource base is being progressively depleted through land degradation of several forms.

At the same time, one-third of the populace remains malnourished. If these malnourished people were to be adequately fed through more efficient and equitable grain distribution, India's famous grain surplus would be instantly eliminated.

Fortunately there is a success story in the state of Kerala. Located in southern India, Kerala is poor in conventional economic terms. But through a fair division of its benefits, it has achieved higher levels of nutrition, health, and education than any other state of India. Its infant mortality rate is only one-third that of India as a whole. The literacy rate, 75 percent, is more than half again as high as in the rest of India. Land reforms have greatly reduced landlessness and inequity.

Bangladesh

In Bangladesh's 55,600 square miles (little larger than New York State or England) live 114 million people, six out of seven of them in the countryside. Feeding such a large populace off such a limited area is a constant challenge. Although the annual growth rate in agricultural production has averaged over 2 percent since 1980, the population growth rate has fallen no lower than 2.4 percent today, meaning that per-capita grain output has been declining.[49] Also because of population growth, the average farm size is projected to decline from 2.25 acres in 1983 to only two-thirds as much in 2005. Because some farms are much larger than the average, well over half of all farms are now smaller than one acre.[50] By the time Bangladesh achieves the two-child family with a total population expected to exceed 250 million, it will have a population density five times that of the densest developed country, the Netherlands, where nine people out of ten live in urban areas, by contrast with only three out of twenty in Bangladesh.

More than four-fifths of Bangladesh amounts to an extended delta at the confluence of one of the largest river systems in the world, comprising the Ganges, Brahmaputra, and Meghna rivers. Yet less than one-tenth of the rivers' combined drainage lies within Bangladesh. As we shall see in Chapter 13 on environmental refugees, there has been a rapid increase in flooding throughout most of Bangladesh in the last two decades. In August 1987, flood damage eliminated one-quarter of the rice crop and the whole of economic growth for two years.

Many Bangladeshis blame India for the growing scale of these flooding disasters. They protest that India could prevent the floods by partially closing the gates of the Farakka Barrage on the Ganges River just over the frontier. The Indians deny the charge. Bangladeshis also object that by using the Barrage to stem the flow of the Ganges in the dry season, the Indians increase the amount of silt deposited in Bangladesh, and so block drainage channels for the floodwaters; Bangladesh's river courses and valleylands receive almost one-fifth of the global silt total in an area only one-thousandth of Earth's surface. Plainly the Farakka Barrage gives India powerful leverage to control almost the entire dry-season flow of the Ganges into Bangladesh. Bangladesh objects that the reduced flow not only diminishes water for its agriculture, but also causes wells in the southwestern part of the country to dry up or suffer saltwater intrusion from the Bay of Bengal. In any case, reduced amounts of water now reach the Barrage each year due to greater irrigation diversions upstream in the politically potent "Hindi belt" of northern India.[51]

These confrontations are particularly acute in the Ganges River basin. The valleylands in both India and Bangladesh supported only 200 million people in 1950, but they now contain 500 million and the total is projected to double within just another twenty-five years.

As a measure of the threats inherent in the India-Bangladesh confrontation overall, note that at the time of Bangladesh's independence in 1971 more than 10 million land-hungry and food-short people crossed the border into India, provoking bloody clashes. Although some of the migrants were impelled also by religious and political reasons, sheer "people pressure" on a limited land-resource base was a prime component of the migratory impulse.[52] Throughout Bangladesh's nationhood, there have been periodic further incursions into India's neighboring states, where the population density is only one-quarter as much, or less. In the 1983 outbreak of violence, 3000 Bangladeshis were killed. (In the Indian state of Assam, a 1991 census revealed that more than 7 million or one-third of the populace were Bangladeshi immigrants.)

Regional Review

The key factor for the three countries is their capacity to feed their ever-growing numbers—another 141 million projected by the year 2000, followed by a further 595 million by 2025. Even today, the average person receives fewer calories than are usually required for an active adult. The three countries' cereal imports in 1989 totaled 5.4 million tons, plus food aid of 1.9 million tons, or enough to make up one-fifth of the diet of 183 million malnourished people (that is, requiring 450 pounds per person per year, but currently receiving only 360 pounds).

Scope for India and Bangladesh to purchase food from outside sources (Pakistan remains a net food exporter) depends to large extent on the strength of their general economies, burdened as they are with foreign debt: in 1990 India owed $70 billion and Bangladesh $12 billion. While these sums appear slight as compared with those of certain Latin America nations, debt servicing nevertheless absorbed 29 percent of India's export earnings and 25 percent of Bangladesh's.[53]

In light of this, the region's leaders may soon find that food security is harder to achieve than military security—even though a decline of the first can be a more potent source of political instability than a decline of the second. Indeed, military security can usually be attained by simple purchase of arms from abroad, whereas food security depends on a complex of factors that are not so readily supplied, primarily a set of environmental and agricultural measures in conjunction with population-planning measures.

On top of all this, there is the problem of widespread landlessness—a factor that in other countries has often proved a source of civil unrest and disorders, if not insurgencies. As far back as 1981, there were 30 million rural families in the region who neither owned nor leased land.[54] When we add in those families with only one acre or less (not enough to support a family even when intensively farmed), the total of landless and near-landless amounted to almost two-fifths of all rural households. If we assume an average family size of six persons, the 1981 total of people affected was roughly as many as the entire populace of the United States.

Overall hangs the prospect of climatic change, with potentially critical impact throughout the region. We shall consider the full implications in Chapter 13 on environmental refugees. Suffice it to mention here that global warming could mean drastic consequences for much of the region. For instance, India's wheatlands in Harayan, Rajasthan, and Punjab, vital to the nation's capacity to feed itself, are at the northern edge of their range. They could not tolerate a temperature increase of more than 2 degrees Fahrenheit, yet global warming in India may well exceed this amount within just a few decades. Nor could the wheatlands "migrate" northwards, since they would run up against the Himalayas. Note, too, that in Bangladesh a climatic disaster in 1974, in the form of broad-scale flooding, caused so much devastation that it led to the violent overthrow of the nation's founder, Sheik Mujibur Rahman.

Security Implications

All in all, then, the region has difficulty in supporting its present 1.2 billion people, let alone another 141 million people (well over half as many as in the United States) by the year 2000. The creditable accomplishment of keeping food production ahead of population growth in much of the region during the period 1960–85 has been due for the most part to mobilizing agrotechnology. During the past eight years, however, there have been steadily diminishing returns to increased technological inputs. Worse, there are abundant signs that these inputs have served to conceal a decline in the natural-resource base in the form of deforestation, soil erosion, salinization, and the like. Yet the resource base is going to encounter still greater burdens as more people with more aspirations impose still greater demands upon it. And all this takes place within a region torn by long-standing religious hostilities.

Over the next few decades we can envision a prospect in which population, land, and food pressures become progressively acute. As a result there may well be attempted migrations on a scale far surpassing what has been the case thus far. In Bangladesh particularly, the combined impacts of population growth, poverty, landlessness, and inadequate food supplies will surely generate pressures for mass migrations into neighboring areas of India, where the population density today is only one-third, and often far less, that of Bangladesh. The result will be a steep escalation of the violent repercussions that have already characterized relations between the two countries. What might be some consequences for security in its proper broad sense? It is not possible here to spell out the precise repercussions. Too many other factors count, such as widespread poverty, inefficient economies, and inequitable social systems, any of which can predispose a nation to instability and thus leave it more susceptible to environmental threats. Also at issue is the lengthy record of strife, confrontation, and periodic hostilities in the region (bearing in mind, too, that two of the nations are now nuclear powers). So might it not be equally appropriate to turn around the basic question and ask, "How can we realistically suppose that environmental problems will not exert a substantial and adverse influence over the prospects for the region's security throughout the foreseeable future?"

We could go one step further. We could ask whether the region's nations could not purchase more real and enduring security by diverting some of their military expenditures toward safeguards for the environmental foundations of their common security overall. There is a military buildup under way on the part of both Pakistan and India, to the extent that Pakistan spent $2.6 billion on military activities in 1987, or 8 percent of GNP, while India spent $9.8 billion, or 4 percent of GNP (Bangladesh spent only $300 million, or 1.7 percent; for comparison, the U.S. military budget in 1987 was $293 billion, or 6.5 percent of GNP[55]). So extensive were these military dispositions that Pakistan ranked 22nd in the world in terms of its military outlays per GNP, but ranked only 116th in terms of GNP per citizen; India ranked 54th and 114th, respectively.

Both Pakistan and India could have made productive environmental use of even a small portion of these military outlays. For instance, $\frac{1}{50}$th would have served to establish tree plantations in 37,000 square miles and 116,000 square miles of national territories, respectively. The Pakistan government believes it urgently needs to reforest at least 37,000 square miles and the Indian government considers that as a national priority its tree cover should be extended by 162,000 square miles. As noted, too, tree planting can serve as a prime strategy for increasing agricultural production—with all that means for food security, political stability, and international equanimity.

Similarly, a small part of military funding could be more productively devoted to population planning. If Pakistan were to achieve replacement-level fertility by 2010 instead of the government-planned target date of 2040, its population, now 125 million, would level off at around 330 million instead of 560 million, a difference equal to almost twice today's population.

Scenarios

A downside scenario foresees that after the failure of the 1994 monsoon, food output plunges. All three countries become major importers of food, even though it hits their meager foreign-exchange reserves. India and Pakistan, having been exporters for years, are sorely embarrassed that they must actually ask for food donations as well as commercial shipments from overseas. Supplies are not sufficient, prices soar, and food riots break out in many cities. India and Pakistan blame each other for instigating the riots on religious grounds. As hunger spreads, more peasants take to felling the last fragments of forests to make way for cultivation—the forests having already been depleted by fuelwood gathering on the part of the region's populations that grow by 23 million people per year. As a result, watershed functions are reduced, leaving less water for the critical irrigation systems: less food again. In response, India and Pakistan announce plans to build a series of dams on their stretches of shared rivers in order to divert extra water. This leads to accusations of water theft, and border skirmishes break out as each side proclaims it is protecting vital interests.

As for Bangladesh, the record typhoon of 1994 drowns one million people, the worst such disaster ever known. It is called a natural disaster, but it is really a population disaster: too many people living in areas they knew would one day be struck by unprecedented calamity. At the same time, the heaviest monsoon on record causes flooding in northern and central territories, too. Virtually the whole country disappears under water. Emergency operations lasting an entire year eat up government funding to an extent that Bangladesh has little left to support agriculture or family planning, let alone investment in other development sectors such as public health. Disease ravages the populace, highways and railroads remain unrepaired, the country collapses. With 40 million people all but homeless if not landless, there starts a mass migration into neighboring states of India. Community clashes multiply, violence flourishes. Pakistan seizes the opportunity to stir up more trouble with India by offering to come to the aid of another Muslim nation.

India decides it is time to teach its main antagonist, Pakistan, a conclusive lesson. The country's military mobilizes and troops cross the border. Pakistan, unable to buy as many arms in recent years because of its massive food purchases, soon finds itself hard pressed. Its agents foment violence on the part of India's minority Muslim communities, and India decides nothing else will do but the toppling of the Pakistan government. Its tanks trundle toward Islamabad. Pakistan retaliates by bombing several mammoth reservoirs in India, causing huge loss of life downstream in the crowded Indian plains and severely cutting India's capacity to feed its people. The war intensifies in every way: it is to be a fight to the finish.

In the finale, mushroom clouds cover the region, including Bangladesh, which is accidentally hit by a series of nuclear blasts. Nobody ever finds out which side released the ultimate strike. Nor is anybody interested—the survivors have a hard enough time surviving. The entire region settles under a cloud of radioactive dust.

An upbeat scenario foresees that in the wake of the 1994 drought, the worst in decades, the nations of the region come to realize they must sort out their water problems: having shared rivers means they will all lose together or all win together. An "Our Water" covenant establishes agreed rights to the entire river flows. The experience encourages them to coordinate on other matters such as Himalayan watersheds and reforestation of upland catchments that, while located in national territories, serve their joint interests. Similar collaborative efforts are undertaken by India and Bangladesh.

All three nations recognize that population growth must be slowed with all dispatch. Pakistan calculates that if it defers replacement fertility by only thirty years, it will end up with an extra 222 million people, or 177 percent of its 1993 population. Its plan to bring down the growth rate from 3.1 percent in 1990 to 2.5 percent in 2000 achieves its goal by 1997. India finally leaves behind the memories of the coercive family-planning campaigns of the late 1970s, and a renewed birth-control effort brings its population growth rate down from 2.0 percent in 1990 to 1.5 percent in 1998. Bangladesh agrees with international agencies that in return for debt-reduction terms, it will embark on a more vigorous family-planning program in conjunction with far greater emphasis on women's health, education, and employment. The Grameen Bank enterprise, run mainly by and for women to support a quarter of a million household-level projects, is replicated by a Green Bank for individuals who engage in environmental activities such as soil conservation and tree planting. The new investment enterprise is so successful that Pakistan and even India send officials to see how it works.

Learning from these common endeavors, the three nations put together the Regional Accords of 1997. India and Pakistan embark on a nuclear disarmament program. All three undertake to collaborate on any activities that affect more than one nation by virtue of the environmental ties that increasingly make rhetoric a reality. An inexorable fact from the start brings peace at the last.

8

El Salvador

If we [the United States] cannot manage in Central America, it will be impossible to convince certain nations in the Persian Gulf and in other places that we know how to manage the global equilibrium.

——Henry Kissinger,
Former U.S. Secretary of State, 1986

El Salvador is the most troubled country in a region about which former president Ronald Reagan stated, "The national security of all the Americas is at stake in Central America."[1] The country has endured more civil strife, political upheaval, military activity, and widespread violence than any other country of the region. It has also suffered more environmental impoverishment than any other country, notably soil erosion, deforestation, and depletion of water supplies.[2] These two concurrent patterns can hardly be coincidental, as was emphasized by a 1982 assessment by the U.S. Agency for International Development: "The fundamental causes of the present conflict are as much environmental as political, stemming from problems of resource distribution in an overcrowded land."[3] Yet so difficult is it for traditional-minded analysts to discern the connections that they were scarcely mentioned in the 1984 Kissinger Commission Report on Central America.[4] Of course, several other factors are at work, such as disparities of wealth and income, inequitable land-tenure systems, and repressive government. But the environmental dimension to El Salvador's predicament deserves far greater attention than it has received.

Within a country the size of Massachusetts live 5.8 million people, half of them farmers. So important is the environmental basis of the economy that two-thirds of export revenues derive from agriculture.[5] Yet soil erosion is so extensive that over three-quarters of national territory is affected seriously or severely.[6] Forests are largely a matter of history. Watershed deterioration is the rule rather than the exception, and water flows from upland catchments are increasingly erratic, with serious harm for irrigation agriculture. As a result of general degradation of agricultural resources, among other factors, El Salvador is increasingly unable to feed itself. Per-capita food production declined sharply throughout the 1980s, and cereal imports increased by one-third.[7]

Agriculture is also beset with an unjust division of farmlands, associated problems of land-tenure systems, and pressures of high population growth. Cropland per person has fallen by three-quarters since 1950, and agricultural advances—agrarian technology, credit systems, and extension services—have not nearly compensated for such a steep falloff. To make matters worse, just 2 percent of the population owns three-fifths of arable land,[8] while half of all farmers are confined to a mere 5 percent of agricultural areas and their average smallholding is only 1.25 acres or less.[9] Almost two-thirds of the farmers are nearly landless or outright landless. As a result, throngs of impoverished peasants are pushed into marginal environments, often steeply sloping areas where the friable volcanic soil erodes easily.

Soil erosion levies a heavy toll on El Salvador's economy. In many areas the landscape now presents a scene of deep gullies and exposed rocks, with no productive capacity short of massive rehabilitation. As a result, rivers dry up for much of the year and groundwater stocks are declining. Only one person in ten has access to clean water for domestic use. Soil erosion also causes sedimentation to reduce the operational lifetimes of major dams, especially the three on the Rio Lempa, which drains half the entire country and contributes almost all of the nation's hydropower as well as much water for irrigation and domestic use. The dams were built to last at least sixty years, but they are not now expected to generate hydroelectricity for more than twenty-five years.[10] Not all the Rio Lempa's system is in El Salvador; its headwaters are located in southern Guatemala and western Honduras, where soil erosion is also widespread and causes political friction with El Salvador.[11]

El Salvador's population density is almost 700 people per square mile. This is between three and eight times higher than that of other Central American countries, and almost as high as India's. Worse, it is projected to exceed 1000 people per square mile by the year 2010. So El Salvador experiences by far the most acute land pressures in the region, even disallowing the skewed distribution of farmlands.

Whether population growth and environmental decline are chiefly at fault[12] or the main problem lies with the unequal distribution of farmlands and unjust division of economic and political power,[13] the resultant pressures have caused a steady migration of Salvadorans into neighboring countries. By the late 1960s, when El Salvador had an average of over 400 people per square mile and Honduras had only 57, one out of eight

Salvadorans had moved into the country next door. Tensions over this migration erupted in 1969 in the so-called Soccer War.[14]

Since 1970 more than 500,000 Salvadorans have migrated to other Central American countries before moving on to countries farther afield, notably the United States. By 1982, at least 500,000 Salvadorans had entered the United States (many of them illegally), or one in nine of the entire population. All told, almost one in four of all Salvadorans, counting internally displaced people as well as international refugees, have fled their homelands.[15] While political repression and inequitable social factors have often played a role, many of these migrants can legitimately be called environmental refugees.

Nor is the future any more promising.[16] Such are the population pressures that even if the government's land distribution and agrarian reform programs begun in 1980 were fully successful, about one-third of the rural poor would still lack secure access to farmland.[17] Today's population of 5.8 million is projected to reach 6 million by the year 2000 and to approach 10 million by 2025 before eventually stabilizing at zero growth with 13 million (Table 8.1). Yet El Salvador could not support more than 10 million people off its own land even if it were to deploy the most advanced and intensive forms of agriculture such as are practiced in Holland and Japan.[18] Already El Salvador produces less than half the food it needs for its fast-growing populace, so it must seek large amounts elsewhere. In 1989 it bought more cereals and received more food aid per person than any other Central American country (except Costa Rica, which could better afford to buy abroad), while the volume of food aid surpassed all other countries combined.

Prospects for purchasing still more food abroad are likewise dubious. Servicing El Salvador's foreign debt consumes one-sixth of export earnings, funds which could otherwise be used to keep the country's graneries stocked. Nor is the economy likely to improve enough to make a difference. Even projecting

TABLE 8.1 CENTRAL AMERICA: BASIC INDICATORS

Country	Population 1992 (millions)	Population Growth Rate 1992 (%)	Projected Population (millions) Year 2000	Projected Population (millions) Year 2025	At Zero Population Growth	Per-Capita GNP 1990, U.S.$ (and average annual growth rate 1965–90, %)		Foreign Debt as % of GNP 1990	Infant Mortality 1991 (per 1000 live births) (U.S. = 10)	Average Annual Growth Rate of Agricultural Output 1980–90 (%)
Guatemala	9.7	3.1	12	22	33	900	(0.7)	37.5	62	2.6
Honduras	5.5	3.2	7	12	18	590	(0.5)	140.9	48	1.8
El Salvador	5.6	2.9	6	10	13	1100	(−0.4)	40.4	55	−0.7
Nicaragua	4.1	3.1	5	8	14	NA*	(NA)	NA	62	−2.6
Costa Rica	3.2	2.4	3	6	6	1910	(1.4)	69.9	77	3.2

*NA = not available.

Sources: Population Reference Bureau, 1992 World Population Data Sheet (Washington, D.C.: Population Reference Bureau, 1992); World Bank, World Development Report 1992 (Washington, D.C.: World Bank, 1992); World Resources Institute, World Resources 1992–93 (Washington, D.C.: World Resources Institute, 1992).

a highly optimistic economic growth rate of 3.9 percent per year, the World Bank does not expect Salvadorans to enjoy the standard of living they had in 1979 (before the outbreak of the civil war) until the year 2006.[19] With a more realistic growth rate for the economy of between 3 and 3.5 percent per year, per-capita income would continue to decline, never to regain its meager 1979 level.

The economic debacle is forcing large numbers of people to return to a state of semi-subsistence in which their main option is to exploit the scarce natural resources available. To make matters worse, half or more of new entrants to the job market within another dozen years are likely to remain unemployed, thus adding further pressures to the natural-resource base if they feel obliged to join the multitudes already practicing a semi-subsistence life-style in marginal lands. Such economic dislocation could only exacerbate El Salvador's instability, ultimately fueling pressures for radical political change.

In the meantime, the country spent extravagant amounts on military activities for most of the 1980s, more than one-quarter of all government spending and one of the highest proportions in the world. The government felt obliged to do so in order to hold down dissident factions within the impoverished peasantry, together with other disaffected communities. The civil war of the 1980s killed 75,000 people, or well over one in every one hundred Salvadorans—a slaughter that has been matched in hardly any other country during the same period.

Fortunately there are two good-news items here. By the mid-1980s the government realized that more children were dying from vaccinable diseases than were being killed or injured from more than $100 million worth of bullets, bombs, and shells each year. So a brief truce was called for a mass immunization program, after which there was a greater proportion of immunized one-year-olds in El Salvador than in New York.

Second, a peace of sorts has broken out. Or at least a peace agreement has been signed. But however much this step is to be hailed with relief, we must recognize that war against the people has been overtaken by war against the land—a war that promises to be more intensely and bitterly fought, and with less prospect of an eventual truce, than the civil war. Fortunately the new president, Alfredo Cristiani, recognizes the vital role played by his country's environments: "Our natural resources are the base of our economic productivity and social wellbeing." But El Salvador, racked for long years by conventional violence, must continue to suffer the travails of ravaged environments and a bankrupt economy, travails increased by the costs of past military conflict.

War outlays expanded fourfold between 1975 and 1990. During the same period, spending on health and education declined by one-fifth. In 1986, one of the peak years of the war, the government spent $211 million on military activities. Just one-fifth of that amount could have done a great deal to shore up the country's environmental foundations. For example, it could have reforested 400 square miles of land, or one-twentieth of the country. Such reforestation would have begun to rehabilitate watershed functions and conserve topsoil, thus enhancing the agricultural resource stocks that sustain much of the economy and populace, notably the peasants who depend on marginal lands and who supplied much of the support for the insurgents. To rebuild the country will cost well over $1 billion, almost half as much as the war consumed.

The military spending record also reveals much of the United States' stance toward the security concerns it perceives in El Salvador. In 1986, the peak year of the war, the U.S. government supplied El Salvador with $122 million in military aid, six times more than the U.S. government spent on environmental measures for the whole of Latin America. Yet a strong case can be made that an environmentally impoverished country will go on experiencing the very economic and social problems, followed by the political upheavals, that the United

States sought to contain through its conventional support for El Salvador.[20]

The long-term outlook for El Salvador cannot but be bleak. As more people try to sustain themselves from environments in ruins, there will be growing discontent and frustration among the bulk of the populace, which is caught in a tightening poverty spiral.[21] The resource in shortest supply will surely be their patience. Increasingly they will feel they have nothing to lose by expressing their grievances through uprisings against the government. In turn, and if the government were to resume its response pattern of the past decades, the upshot would be another spiral, this one of repression fostering still greater resentment, triggering more repression again. It is realistic to suppose that El Salvador's future could continue to be shrouded with the smoke of escalating conflict.

Other Countries of Central America

The environmental travails and pervasive poverty of El Salvador are replicated, though not to such an extreme degree, in most of the rest of Central America, notably in Guatemala, Honduras, Nicaragua, and Costa Rica. Of these, Guatemala, Honduras, and Nicaragua rank among the most violence-racked countries on Earth. Indeed, most of the region features widespread discontent, with periodic outbursts of violence.[22] Worse, the potential for future trouble is great and growing: inadequate food supplies, economic deficiencies, unemployment, broad-scale migration both within and among countries, civil disorders, and political turmoil.

As in El Salvador, many of these problems are rooted in environmental factors. Income derives mainly from natural resources that support agriculture, forestry, fisheries, and hydropower energy. More than half the region's economies are based upon natural resources, which also account for two-thirds of all employment and most export revenues.[23] So environmental health is vital to the region's economic wellbeing, and in turn to its political stability. Yet today it is facing an environmental crisis unparalleled in its history.[24]

Much of the region features mountainous terrain with friable volcanic soils, hence they are highly vulnerable to soil erosion. In the case of Guatemala, as much as two-thirds of the national territory is at risk. At the same time, the region's forests have declined by two-thirds since 1950, and within another decade there could be little left beyond isolated fragments. Deforestation serves to accelerate soil erosion, which now affects over half the region's agricultural lands, leading to a decline in soil fertility and crop production.[25] Food output per head slumped during the 1980s by over one-tenth in Honduras and Costa Rica, by one-fifth in Guatemala, and by one-third in Nicaragua.[26] Food imports soared by three-quarters. For details of some basic indicators of agriculture, among other forms of human welfare, in these countries, see Table 8.1.

Soil erosion also reduces hydropower generation. In Guatemala, most watersheds contributing the bulk of hydropower are in an advanced state of degradation. In Costa Rica, virtually every major hydropower facility suffers from sedimentation, with somewhere between $115 million and $274 million of electricity forfeited in the space of two decades at the Cashi Dam alone.[27] Costa Rica used to get all its electricity from hydropower with a surplus exported to Nicaragua, but today the export has vanished and the country must gain one-tenth of its electricity from diesel-burning turbines, increasing its oil-import bill.

On top of all this is population growth. The aggregate population of the region, apart from El Salvador (but including Panama, not counted as part of the political region), has grown from 6 million in 1950 to 25 million today (Table 8.1) and is projected to soar to 48 million in 2025. If other things were equal, this problem alone would have led to fast-growing pressures on the natural-resource base supporting the region's economies. But other things are far from equal. The division of farmlands is heavily tilted in favor of a few wealthy landowners. In Guatemala, for instance, 2 percent of farmers control 80 percent of croplands, while 88 percent subsist off plots too small to support a household. In Honduras, 5 percent of landholdings account for 60 percent of arable lands; in Costa Rica, the numbers are 3 percent and 54 percent, respectively; and in Nicaragua, half of farms occupy more than nine-tenths of farmlands, while the other half account for less than one-twentieth.[28] As a result, well over one-quarter of the rural populations in Guatemala and Costa Rica are landless, and almost one-third in Honduras.[29] This grossly unequal distribution causes small-scale farmers to overwork their diminutive holdings and thus to aggravate the falloff in soil fertility, as well as to migrate into marginal environments.

The plight of the peasantry has converted many countryside communities into natural allies of insurgent groups throughout the region. As a result, the flames of revolt and war have spread far and wide, making this one of the most insecure regions on Earth.[30] As a further result, there has been an increase in repressive government and in military responses to social disorders. Guatemala, for instance, now has two and a half times as many soldiers as in 1978.[31] Moreover, there has been a massive surge in the number of refugees, whether within countries or across national frontiers. During the 1980s as many as 2 million Central Americans became uprooted.[32] This amounted to one person in eleven, equivalent in the United States to the entire populaces of New York and New Jersey. Of these, half have made their way to the United States. Given the population growth and the environmental rundown foreseen, the next few decades will surely see many more millions seeking sanctuary across the Rio Grande, with all the economic and social dislocation this could entail for the United States.[33] I shall return to this aspect in the next chapter on Mexico.

The Experience of Haiti

As a portent of what could lie ahead, consider the case of Haiti. This is a country of the Caribbean rather than Central America, but it likewise suffers much environmental decay, it has long been beset by political disorders, and it is close enough to the United States to have encouraged many of its inhabitants to flee there.

Whereas almost all of Haiti was originally forested, only 2 percent remains so today.[34] In consequence, there has been widespread soil erosion; in some localities, half the landscape is bare rock.[35] Almost three-quarters of today's populace of 6.5 million depend directly on agriculture. But two-thirds of farmlands have slopes of more than 20 degrees, that is, they are super-sloping, and they support farmers at a density of over 700 to a square mile (as many as the most densely populated areas of India), so they are especially prone to erosion.[36] Between 1950 and 1990 the amount of productive arable land declined, through erosion and other soil degradation, by well over two-fifths. Per-capita grain production is now only a little over half what it was forty years ago.[37] At the same time, deforestation has reduced stream flows and irrigation capacity. The Avezac Irrigation System, planned to cover 9500 acres, now supports only half as much. The sediment-clogged Peligre Dam, representing half the nation's hydropower potential, will operate only half as long as originally expected.[38]

Fortunately there is a slightly brighter side to the picture. Due to deforestation, it has been projected that most stocks of fuelwood—the chief source of energy for most households—would be finished by the mid-1990s. The situation has been largely ignored by the government, which has launched only a handful of reforestation projects. Enter an alliance of nongovernmental bodies, made up of American relief agencies such as

CARE, Oxfam USA, and Catholic Relief Services working in conjunction with scores of counterpart groups in Haiti. By 1990, they had enabled 135,000 farmers to plant 35 million trees. Many farmers each planted as many as 300 trees on their smallholdings. True, this is only a small start on a solution. But Haiti can use all the starts it can get.

Meanwhile, Haiti's troubles leave it the poorest nation in the Western Hemisphere and one of the most destitute nations anywhere.[39] Small wonder that at least one million Haitians, or over one in seven of the populace, have reportedly left their homeland, most of them illegally entering other Caribbean countries.[40] The bulk of these environmental refugees (some political refugees, too) derive from northwestern Haiti, where deforestation is worst, nine-tenths of all farms are smaller than three acres, and four out of five children suffer from malnutrition. As far back as 1981, 122,000 migrants (possibly a good many more according to some observers) had made their way to the United States.[41] The most desperate of these migrants reached the United States via the hazardous 600-mile boat trip to southern Florida, where an estimated 24,000 reside today, with another 56,000 in other southeastern states.[42] To support these new arrivals, the Florida state government has been spending more money than the U.S. government has been allocating to foreign aid for Haiti. Many more will surely be on their way in the years ahead. In the single month of April 1992 the U.S. Coast Guard picked up 10,000 Haitians.

This last factor highlights a salient aspect of American foreign policy. If the United States does not do more to safeguard the environments and to support the economies of Caribbean countries, will it not eventually encounter sizable concealed costs of inaction?[43] To prevent the unemployed and impoverished total from rising above the 1980 level, Caribbean countries will need to create 11 million new jobs at an annual cost of $14 billion a year over a period of thirty years.[44] The present U.S. aid to the Caribbean is less than $1 billion a year. Would it not be a sound investment for the United States, together with other developed nations, to protect its own interests by greatly expanding its aid support? Expensive as this might sound, it will be far less costly than the price Americans might eventually find themselves having to pay—and the dollar price could turn out to be less than the entire price if floods of destitute migrants cause social dislocations.

Even more productive would be to expand the opportunity for Caribbean exports to find markets in the United States, despite Washington's push toward trade protectionism. After all, and this has been repeatedly demonstrated by the Commission for the Study of International Migration and Cooperative Economic Development under Congress, trade liberalization would foster trade in commodities rather than migrants. Much the same considerations apply, of course, to Central America[45] (also to Mexico, as we shall see in the next chapter).

A Better-News Item: Costa Rica

Fortunately in Central America there is a partial exception to the pattern of poverty, turmoil, and conflict. Costa Rica long ago disbanded its army, and this has freed up government funds for education and health among other development sectors (today there are more teachers than soldiers and police combined). The country has enjoyed a series of environmentally minded presidents. The annual growth rate in agriculture in the 1980s exceeded the population growth rate, the only country of the region to achieve that. Life expectancy averages seventy-five years, a figure comparable to most developed nations. Education is free and compulsory for all. Similarly exceptional are nutrition, sanitation, and basic health care. Women enjoy extensive rights and opportunities. Family-planning services are available everywhere. The population growth rate has plunged from 3.8 percent in 1960 to 2.4 percent today—though this remarkable family-planning record has recently stalled in parallel with economic stagnation.

Regrettably, a growth rate of 2.4 percent still means the population would double in another twenty-nine years. Despite Costa Rica's many fine efforts to safeguard its environments, it is difficult to see how the country could accept twice as many people within such a short time.[46] During the 1980s the agricultural frontier closed as farming settlements spread to both oceans and both borders. For the first time in more than four centuries of post-Colombian history, and with three and a half times as many people as in 1950, Costa Ricans have no ready access to new land. This predominantly agrarian society is having to adjust to a sudden change from land abundance to land scarcity.[47]

As we have seen, moreover, the natural-resource base is growing less and less capable of supporting the present populace. Much of the best soil cover has gone. Electricity demand is growing by 8 percent annually, yet hydropower capacity is waning in the wake of deforestation and soil erosion. There are mounting pressures for water supplies for irrigation, industry, and domestic needs. Pollution problems are growing worse faster than ever. By 1995 it is likely there will be no more commercial forests left. All this will surely lead to a spread of poverty and hence to a possible resurgence in the birth rate. At the same time, the country is saddled with foreign debt worth almost $1000 per person, or over half as much as GNP per head.

Even in this country, a democratic model for Central America, the outlook is far from bright. And the consequences will not be confined to Costa Rica or, in the case of other countries of the region, to Central America. Apart from the political ramifications of persistent instability south of the Rio Grande, there are ecological linkages to Americans' wellbeing. Central America's forests are considered by the National Cancer Institute outside Washington, D.C., to harbor an unusual proportion of plant species with potential against cancer; as a result of deforestation, we are certainly losing several of these species every year. Even more to the point, deforestation deprives North American bird migrants of their wintering grounds. Each spring, returning migrants of dozens of species reveal declining populations[48]—and they reach the United States at the time when many insect species appear again, their numbers kept in check by the birds. Without these natural controls of insects, American agriculture could eventually start to find trouble in a different sort of silent spring.

Scenarios

A negative scenario foresees that the new democratic regime in El Salvador fails to surmount the appalling problems, both environmental and economic, that are the legacy of decades of misrule, population pressures, and environmental ruin. A new insurgency erupts in 1995, receiving aid from Cuba. Civil war resumes on a scale greater than ever, with the bulk of the rural population, despairing of any improvement in their lot, supporting the rebels. Within months, more than one million Salvadorans seek sanctuary in neighboring countries. But the hoped-for hosts reject the idea flat. They have their own mounting crises of deforestation, eroded farmlands, water shortages, and the like, plus population pressures from almost twice as many people within a generation. They invade El Salvador to "restore order," albeit order of a type that accords with their generally authoritarian and repressive regimes. Dissident groups in these countries, seeing their governments and militias occupied elsewhere, take their chance to foment revolution in their own lands. Several parts of Central America are ablaze, and the violence promises to persist indefinitely.

A positive scenario anticipates that El Salvador wearies of its strife and no-hope economy. A coalition of all main political groups, plus special-interest communities such as the peasantry, puts together a program of rebuild and renew. The United States supplies one dollar for economic restoration to match each dollar it earlier provided for military destruction. Central to this U.S. initiative is a countrywide plan for environmental recovery. Under Trees for Tomorrow, soldiers reforest a whole one-tenth of the country within the first two years. Success stories apply also to soil and water conservation. These measures are reinforced by agrarian reform (sharing out of farm holdings in an equitable manner), which soon proves to be the biggest breakthrough on that front since Taiwan forty years ago. Rejuvenated croplands produce the best harvest since before the civil war. Many El Salvadorans who had earlier fled the country return to their homeland. Salvador City soon possesses twice as many Salvadorans as Los Angeles. Yet this influx of people does not cause a sudden burst of population growth since it is more than offset by a sustained plunge in fertility.

Elsewhere in Central America, reformist groups follow El Salvador's lead. Guatemala and Honduras, after decades of oligarchical and repressive rule, give way to broad-scale democracy. The old elitist and crony-ridden bureaucracies are swept aside and replaced by leadership cadres that the citizenry senses can reflect their basic needs. This political demarche leads to economic freedom as well, releasing entrepreneurial energies on every side. The new community-based governments appreciate that their development future is closely tied up with their environmental future. The Environment for Everybody movement spreads throughout the region.

A 1996 conference of all seven nations acknowledges that environmental facts of life make the region one, however little that has been appreciated by politicians in the past. More important still, the long-sought federalist spirit starts to emerge; the first expressions stem from a series of internation measures to protect transfrontier watersheds. In turn, this triggers a pervasive spirit of cooperation at all levels—regional, national, and local. Rehabilitation of hydropower catchments is undertaken on a communal basis.

These initiatives receive powerful support from the United States, partly out of self-interest in order to stem the flood of migrants across the Rio Grande. It is further buttressed by concessional loans from the Inter-American Bank after Wall Street and the World Bank engage in broad-scope debt relief. Two years later, Costa Rica is not the only country to feature more teachers than police, and the last militias are dismantled as a symbol of an outmoded past. The diversion of military funding into development sectors—smallholder agriculture, forestry, education, public health, and women's needs—triggers an outburst of grass-roots response. Central America grows from the bottom upwards, and with such enthusiasm that it seems the sky's the limit.

9

Mexico

The consequences [for Mexico] of not creating nearly 15 million jobs in the next fifteen years are unthinkable. The youths who do not find them will have only three options: the United States, the streets, or revolution.

—J. G. Castenada,
Professor of Political Science,
National University of Mexico, 1985

Mexico suffers severe environmental strains. Four-fifths of the country is semi-arid or arid, an expanse that is increasing through desertification, soil erosion, and general land degradation. Agricultural lands amount to only 155,000 square miles,[1] the same as the state of California, on which the country must support almost 90 million people today. Despite these unpromising conditions, Mexico managed to feed its people from 1950 to 1984, quadrupling its grain harvest during that short period. But today at least 70 percent of agricultural lands are affected by soil erosion,[2] and a full one-tenth of irrigated lands are highly salinized, with huge areas needing costly rehabilitation if they are to grow food again.[3] Desertification eliminates around 1000 square miles each year.[4] Probably worst of all, global warming with its drought effects threatens to reduce rainfed agriculture by a whopping 40 percent.[5]

Moreover, the country's remaining forests total only 56,000 square miles, and one-twentieth fall each year to large-scale ranchers and small-scale peasants.[6] The deforestation rate means that all forests will disappear in twenty years even if the rate does not increase. The loss of the forests' "sponge effect" disrupts river flows; and in two-thirds of arable lands, water supplies are the main limitation on agriculture. Because of failing water supplies together with soil erosion, at least 400 square miles of farmlands are abandoned each year.[7]

A while ago there seemed to be relief from the problem of too few food-producing lands for too many hungry Mexicans. From the late 1950s to the early 1970s, the area under staple crops—corn, wheat, rice, and beans—expanded by almost half.[8] Unfortunately, much of this expansion had to occur in marginal lands with highly erodible soils[9]; today more than half of all farmers occupy the one-fifth of croplands that are steeply sloped.[10] By the mid-1970s, there emerged a decline in crop harvests in these newly opened-up areas—precisely at the time when there was a peak in Mexico's population growth. During the 1980s, the country has lost one acre in ten of its grain-growing lands, either because the soil has become exhausted or because the land has been taken for other purposes such as urban expansion.[11] Today food production has slipped well below its 1970 level, and at least two people out of five are unable to obtain the minimum amount of calories for health, growth, and a productive life.[12] Mexico now has to import one-quarter of its grain, a financial burden for its troubled economy.[13]

This difficult situation does not reflect environmental problems alone. Other factors make the situation worse. The unequal distribution of agricultural lands is becoming more acute as large farmers buy out small farmers. As a result, growing numbers of peasants impose still greater strains on overworked croplands, reducing their fertility. Thus there is a trend for migrating throngs of small farmers to be pushed into marginal lands, precisely the lands most susceptible to degradation.[14]

As a result of this agricultural squeeze, there has been a recent strong upsurge in migration from Mexico's rural areas.[15] Many migrants make for Mexico's cities, which, after growing by 5 percent or more per year since 1960, are less and less capable of absorbing more arrivals. For every two rural Mexicans who head for the city, one now crosses the border into the United States, partly drawn by the prospect of a better life in the United States and partly pushed by poor and deteriorating conditions in Mexico. Push becomes a function of pull, and vice versa.[16] We shall return to this migration topic later in the chapter.

Hardly any other city reveals the problems of over-rapid urbanization more than Mexico City. It has mushroomed from 9 million people in 1970 to almost 20 million today, or more than the combined populations of Sweden, Norway, and Finland. It is projected to keep on growing to 25 million in the year 2000 or very shortly thereafter. The proportion of citizens enjoying clean drinking water and adequate sanitation has been declining. The city features half of the country's industry, and it spews out 3.3 tons of contaminants for each of its residents. The average lead level in residents' blood is four times that of Tokyo's citizens, and breathing the polluted air is reckoned as bad as smoking two packs of cigarettes a day.[17] So acute is the situation that the government has decreed that car owners shall be banned on a rotational basis from driving one day a week—though given the growth of the city's car population, this measure will not help to reduce pollution for more than a brief interim.

In these inauspicious circumstances, it is not surprising that certain of Mexico City's inhabitants have taken to joining the throngs of rural residents heading off in search of a more tolerable life in the United States. This recent trend swells the migratory flood.

Population

As for what is probably the biggest problem of all, population growth, the outlook must be considered bleak insofar as Mexico cannot adequately support (feed, employ, house, generally care for) its present populace of almost 90 million people, let alone ever greater numbers. Fortunately there has been some success in the family-planning field. As recently as the early 1970s, Mexico's population was growing as fast as virtually any in the world, at 3.2 percent per year. Then came a dramatic turnaround, partly because almost all females were receiving some schooling and partly because the infant mortality rate plunged. Thanks to a broad-scale birth-control program, the government found that between 1972 and 1984 every peso spent on family planning saved nine pesos on mother and infant care, for a total saving of 318 billion pesos or more than $2 billion.[18]

As a result, the average family dropped from almost seven children to fewer than five in 1980 and below four in 1990. Today the growth rate is down to 2.3 percent per year.[19] Nonetheless, and because of the youthful profile of the populace and hence the built-in "demographic momentum," the year 2000 could well see a projected total of 107 million people, surging to 143 million by the year 2025.

Of the 31 million people in the current labor force, at least 15 million are considered unemployed or underemployed.[20] The total number of unemployed or underemployed workers is expected to exceed 20 million by the year 2000.[21] The work shortage is worst in rural areas, where unemployment and underemployment affect two out of three of all would-be workers—and where three out of five farming households are near-landless or outright landless.[22]

Mexico's Future

What are the prospects for Mexico's future, given the twin pressures of population growth and environmental decay? Despite an exceptional performance for much of the period 1960–80, the economy has generally stagnated or even contracted since 1980 while the population has grown by more than one-quarter. Real wages have been in free-fall for years; in 1988 they were 40 percent below those of just six years earlier.[23] Two out of five workers now receive less than the official minimum wage, insufficient to meet basic needs. Around half the population does not obtain a suitable supply of calories, and well over one-quarter are chronically malnourished.[24] The per-capita GNP, $2490, is among the highest of the major countries of Latin America, but the bottom 20 percent of the population enjoys only 3 percent of national income. Mexico is a richish country full of poor people.

Nor does the economic outlook presage much improvement.[25] Other troubles apart, Mexico suffers from a bad dose of foreign debt. Outstanding debt totals $97 billion, and Mexico is paying interest at a rate twice as large in proportion to its GNP as did Weimar Germany (a burden that eventually led to Hitler's revolution). A realistic prognosis is that the average Mexican in the year 2000 is unlikely to be better off, and could well be poorer, than today.[26] Those citizens not interested in averages, the destitutes at the bottom of the pile, could be experiencing poverty to match today's but in far larger numbers. In turn, this spread of poverty could reverse the recent success in family planning (impoverished people tend to have the most children) and trigger a renewed upsurge in the population growth rate.

Moreover, to keep pace with the growing work force, Mexico

will have to create one million new jobs a year, this being a full one-half the rate the United States achieves, while doing it off an economy only one-thirtieth the size, or about two-thirds that of Greater Los Angeles. Generating these jobs will require an investment of as much as $500 billion.[27] (In 1986, a year with poor economic performance due to declining oil prices and growing debt burden, the jobs total actually decreased.) Not surprisingly, it is an optimistic estimate that foresees "only" 20 million Mexicans (could be many more) without adequate employment in the year 2000—and at least half of these will be living in rural areas, where there could be an additional 2 million, perhaps even 4 million, landless peasants.

Migration
to the United States

This all makes for a sizable pool of potential migrants to the United States.[28] Mexico has a 2000-mile border with its giant neighbor, the longest such border between a developing country and a developed country. Across the border comes a flood of Mexican migrants—and growing troubles in Mexico, whether economic, environmental, or demographic, could trigger wave upon surging wave of additional migrants. Right now there are at least 150,000 legal migrants a year, with another 150,000 to 350,000 (possibly many more) illegal migrants, all of whom travel to stay (still larger numbers cross the border temporarily). The cumulative total of both legals and illegals is now 3 million, possibly 4 million—and growing rapidly. Within just a few more years the number could readily top 10 million, with little sign of an end in sight to the migratory surge.[29]

Whether the overall ultimate impact on American society will be positive or negative is hard to say in the long run. It will surely be significant.[30] Already many Americans, fearing job losses and cultural disruptions, want the flow stemmed if not stopped. Many Mexicans, fearing for their hard-pressed economy and population growth, want the migration maintained if not expanded. Were this "safety valve" to be eventually restricted through tighter U.S. controls, the result could well be an increase in social turmoil and political instability in Mexico. On the other hand, recent migration trends could persist with immigrant numbers swelling year by year. Counting other arrivals apart from Mexicans and other Hispanics, legal immigrants now total well over one million a year, with illegals making up at least as many again.[31] The legal immigrants cost the United States $3 billion a year, or $3000 each, for resettlement services. An immigrant flood of this order could eventually carry profound consequences for the very concept of what it is to be American. Within twenty years the United States population would climb from 258 million to over 300 million, and within another three decades it would top 350 million.[32] The bulk of immigrants would likely be, as today, Hispanics from Mexico and Central America (see Chapter 8)—and the proportion seems set to swell as the years go by.

This means the ethnic makeup of the United States would shift profoundly. Within the lifetimes of most Americans now alive, Hispanics would constitute one in five of all residents; they would have become more numerous than blacks and hence the largest minority within less than two decades, by 2010.[33] As a community, Hispanics tend to be downwardly mobile. When they set foot into Uncle Sam's melting pot, moreover, they refuse to be melted, preferring to retain their own Spanish language. In turn, this means that in parts of Florida, Texas, and California, those who do not speak Spanish find it hard to get a job in restaurants and other places where Spanish is essential, which is tough on those educationally disadvantaged, primarily blacks. The result is already here— flare-ups across several of the border states.

Mexico's doubtful prospect is important for U.S. interests in still other ways. The country is the third largest trading partner of the United States—though its enduring economic crisis and austerity program during the 1980s led to reduced imports, costing 500,000 American jobs by 1987.[34] American banks still hold around $40 billion of outstanding debt in Mexico, which further depresses the Mexican economy; to this extent, blue-collar Main Street has been forfeiting jobs to permit debt payments to white-collar Wall Street. Were Mexico to default on its debt, the backlash would reverberate throughout the American banking system. In the long run, moreover, persistent economic stresses could lead to political upheavals of a scope to exceed those the United States now grapples with in Central America.

This all helps to build a plausible prospect of persistent instability and insecurity in Mexico. It is realistic rather than alarmist to envision a not impossible scenario: revolution. After all, the makings of revolution—declining living standards, rising unemployment, soaring aspirations—have all been gathering force for well over a decade. In a revolutionary outcome, the outflows of Mexicans already seeking economic survival in the United States would be joined by those seeking physical safety. Recall that the 1910 revolution impelled at least half a million out of 15 million Mexicans to seek sanctuary north of the border. Recall, too, that in that era the great majority of Mexicans lived far away from the border, whereas today 6 million are within ten miles of it,[35] and tens of millions more have heard that passage to the United States is hardly difficult.

Scenarios

On the downside, Mexico's troubles deepen on every side. Its economy cannot keep up with all the demands on it, such as expanding poverty, low oil prices, and foreign debt. The decline in family size, which had been a success story for over a decade, comes to a halt as people feel they can no longer believe in their economic future: the prospect is demographic doomsday. Mexico City's living conditions become so intolerable, especially after the 1994 smog disaster with its pandemic of respiratory diseases, that an exodus begins toward the countryside, notably toward the only unoccupied lands remaining, arid areas and forests. Environmental degradation accelerates, soil erosion reaches disastrous proportions, and extensive areas become desertified. Global warming takes hold, whereupon Mexico becomes warmer and drier, a situation made worse through reduced rainfall. The result is that non-irrigated croplands produce less than half, sometimes only one-fifth, of what they had before.[36] By 1997 the outlook looks beyond remedy, and the migratory surge across the Rio Grande swells several times beyond earlier levels, reaching 5 million in 1998. In reaction, the American government tries to close most of the border. This "cactus curtain" simply triggers fresh desperation in Mexico, and still larger multitudes seek sanctuary in the United States with enraged throngs storming border posts. Relations between the two nations deteriorate fast until there is talk of frontier hostilities.

The overall situation slumps to a point that engenders what some people felt was long an inevitable outcome: revolution. The new government is socialist inclined and distinctly undemocratic. The United States intervenes on the alleged grounds of protecting vital interests, and this leads to outright confrontation between the two countries.

An upbeat scenario foresees that Mexico stabilizes its economy, thanks in part to increased oil revenues as industrialized nations charge consumers much more for fossil fuel, a step that is reciprocated by steady price hikes by oil producers. Industry and high-tech manufacturing take off, and the long-standing economic slowdown turns into a speedup. With its newfound wealth, Mexico finds it can handle its debt burden better, and expanded payments are counterbalanced by major rescheduling on the part of American banks, which finally realize their best bet lies with a Mexican economy that is growing rather than stagnating. The resurgent economy supplies better living conditions to Mexico's poorest people, and the birth rate plunges.

In addition, an enlightened government—the one that gained plaudits from conservationists with its far-reaching 1992 Biodiversity Plan—turns its attention to agriculture, especially subsistence farmers who have been bypassed by the Green Revolution and cause much environmental damage. A Gene Revolution enables small-scale farmers to engage in organic agriculture with low inputs and medium-level outputs. Now that this peasant sector is mobilized in commercial rather than subsistence agriculture, and the energies of millions of smallholders are released, there is an upsurge in grain harvests. By 1995, Mexico becomes a grain exporter again. Even better, soil erosion declines as peasants find they have less need to expand into marginal environments.

There is a sizable contribution from outside too. Particularly helpful is the massive American investment in Mexico's economy, partly as a result of the North American trade agreement embracing all three countries, partly as a result of Congress's imaginative supply of incentives for those entrepreneurs who want to engage in the Broadened Border program. Under this program, the aim is to build an expanded frontier zone that extends for dozens of miles on each side, based primarily on American capital and Mexican labor. This area, roughly 100,000 square miles, becomes a land that, far from being a no-man's land, is an everyone's land, albeit of undeclared sort. It serves to absorb most of the workless and landless Mexicans who would otherwise become migrants into the entire United States. The Rio Grande Hands-Above-the-Frontier breakthrough, as it becomes known, starts to serve as a model for other parts of the world with migrant problems.

CASE
STUDIES:
GLOBAL
EXAMPLES

10

Population

Short of nuclear war itself, population growth is the gravest issue the world faces. If we do not act, the problem will be solved by famine, riots, insurrection, and war.

—Robert McNamara,
Former President of the World Bank

If we do not get population growth under control, habitat on Earth will be destroyed by ecological disaster and/or violent migration processes.

—Willy Brandt,
Former Chancellor of
the Federal Republic of Germany

I have been to many population conferences over the years, also a number of specifically security conferences. Each has had much to say on its own subject, but strangely enough there has been all too little attention directed at their connections. In this chapter we shall consider how far there are in fact some linkages.

There is plenty of evidence. We have seen that part of the standoff between India and Bangladesh stems from sheer pressure of human numbers on both sides of the border. The Soccer War between El Salvador and Honduras erupted when throngs from overpopulated El Salvador flocked across the border into "empty" Honduras. The traumas of Ethiopia would have been far less likely to arise if the population had not increased by over half in just twenty years and thus reduced the environment's capacity to support fast-growing numbers. As this book has made plain, there are many other such instances.

In short, environmental problems are compounded by the factor of population growth, if not caused by it. This factor serves both to exacerbate environmental decline and to leave still larger numbers of people suffering environmental impoverishment. Thus there is great scope in population growth for conflict of multiple types—scope that will increase as growing numbers of people try to sustain themselves off declining environments.[1]

Population growth generates an entire series of adverse consequences. Environments are misused and overused. Per-capita economic advance is slowed and per-capita food production stagnates. Social services are overburdened. Development generally proceeds at a slower pace. And this is to put the situation at its best.[2] Instances include most countries of sub-Saharan Africa and southern Asia, these being the countries with the worst track records in development and with the highest population growth rates. A number of other countries, notably in Central America and along the Andes, display a development performance that would surely be improved if they did not have to cope with rapidly increasing numbers of people.

Consider again the case of Ethiopia. The deforestation and soil erosion in the highlands during the 1960s and early 1970s was due in major measure to pressures of a population that increased from 16 million in 1950 to 25 million in 1970, an expansion of more than half in just twenty years. This clearly contributed to the environmental debacle of the early 1970s and the overthrow of Emperor Haile Selassie. Today's security outlook in Ethiopia must be reckoned still more unstable with a population that has grown to 56 million—and is projected to reach 71 million in 2000 and 140 million in 2025.

But if there is indeed a connection between population and conflict, how does it work? What is its "operational chemistry"? Do population problems directly and inevitably lead to violence? Or do they work indirectly, for example, in catalytic conjunction with other factors such as environmental decline? If the latter, does the "other-factor" complication make population itself less potent as a source of conflict? Or does it make it all the more dangerous, in that population pressures then work in less overt, and hence less heeded, fashion?

Population Numbers

We are in the middle of an unprecedented expansion of human numbers. It took 10,000 lifetimes for the world's population to reach 2 billion people. Now in the course of a single lifetime, it is increasing from 2 billion to three times as many, and within another lifetime it could well double again.

Global population in 1992 amounted to almost 5.5 billion people. Of these, 4.2 billion, well over three-quarters, were in developing nations and 1.2 billion in developed nations[3] (Table 10.1). The total was growing at a rate of 1.7 percent per year, or 93 million people (the equivalent of a "new Mexico"). The rate of annual increase will not peak until 1998, with 98 million people. As we shall see below, it is this factor—the annual increase in absolute numbers—that is also critical to the population prospect. Of the annual increase, more than 90 percent is in developing nations, and over half in Africa and southern Asia, which feature the majority of the 1.2 billion "poorest of the poor." By definition, developing nations have limited capacity to cope with the environmental and economic consequences of ultra-rapid growth in human numbers, due to their

TABLE 10.1
WORLD POPULATION BY REGION, 1992–2025
(Millions)

Region	1992	Projected 2000	2025	Annual Growth Rate 1992 (%)
Developed Countries	1224	1274	1392	0.5
Developing Countries	4196	5018	7153	2.0
Africa	654	884	1540	3.0
(Sub-Saharan Africa	526	668	1229	3.0)
Asia	3207	3718	4998	1.8
Latin America/Caribbean	453	535	729	2.1
North America	283	298	363	0.8
Europe	511	515	516	0.2
Oceania	28	31	39	1.2
World	5420	6292	8545	1.7

Source: Population Reference Bureau, *World Population Data Sheet 1992* (Washington, D.C.: Population Reference Bureau, 1992).

low per-capita incomes, unproductive agriculture, meager technology, and inadequate investment all around. Even if they had all these requisites, they would still be hard pressed. A country as organized as Switzerland with its clocklike precision of planning could hardly take on board twice as many people in just a generation.

The latest projections[4] indicate that the global total will reach 6.3 billion people by the year 2000 and 8.6 billion by 2025, before leveling out at a total of 11.3 billion by 2100 or shortly thereafter. The 11.3 billion figure for the 2100 global total is the medium-level projection. The high projection indicates the total could attain 14.2 billion (or even 20.7 billion), while the low projection indicates it could be held as low as 7.9 billion (and even 5.6 billion). The difference between the high and low projections for 2100, 6.3 billion, is much more than today's total of 5.5 billion, and thus serves as a graphic measure of the maneuvering room still available.

Especially significant is the situation of 1.2 billion absolutely impoverished people in developing nations. They are people who most need the benefits of development, who often cause a disproportionate amount of environmental damage, and who have the highest fertility rates.[5] While their present one-fifth share of total numbers is likely to decline, their absolute numbers are projected to keep on increasing for several decades.[6]

Note that all these demographic projections are no more than extrapolations of recent trends, together with some assumptions about the role of economic advancement and family planning in stemming the rate of population growth. Projections are not predictions, and still less are they forecasts: demography is not destiny. By their nature, they take no account of other variables such as policy changes and technological breakthroughs. Nor do they reflect environmental factors: they assume there will be no Malthusian constraints. But as we shall see below, the rapid degradation of the environmental base that ultimately supports all communities could soon start to exert a marked effect on population growth. Given the record of the last two decades, it becomes increasingly hard to see how sub-Saharan Africa, for example, will experience a projected quintupling of human numbers within another century as long as gross environmental impoverishment continues to spread. Per-capita incomes are generally no higher than in 1960, and per-capita food supply is one-fifth less. Those who consider that population growth may soon be pressing against or even exceed "carrying capacity" (an important concept to be considered shortly) are inclined to be skeptical about demographic projections made in an "environmental vacuum."

Much economic growth of the recent past is clearly not sustainable for environmental reasons alone.[7] This, in conjunction with the linkages between environmental degradation and population growth, alters the outlook profoundly.

It may well turn out that we have achieved economic advancement in the past at environmental cost to the future's potential for still more advancement—and even at the more serious cost of an actual decline in human welfare. Consider the case of Green Revolution agriculture, which enabled growth in grain production to keep ahead of growth in human numbers throughout the period 1950 to 1984. There appear to have been certain covert costs in the form of overloading of croplands leading to soil erosion, depletion of natural nutrients, and salinization of irrigation systems. These costs, while unnoticed or disregarded for decades, are now levying a price in terms of cropland productivity. In Pakistan one-fifth of irrigated lands, and in India one-third, are so salinized that they have lost much of their fertility. Yet these two countries have often been proclaimed as prime exponents of Green Revolution agriculture.

In fact, environmental problems are now causing sizable cutbacks in food production at a time when population growth continues to soar. Soil erosion leads to an annual loss in grain output estimated at 9 million tons; other problems such as salinization and waterlogging of irrigated lands are responsible for 3 million tons lost. In addition, there are various types of pollution damage to crops, worth another 2 million tons.[8] So the total from all forms of environmental degradation comes to 14 million tons of grain output per year. This total is to be compared with gains from increased investments in irrigation, fertilizer, and other inputs, worth 29 million tons per year.

Thus environmental factors are now causing the loss of almost half of all gains from technology-based and other advances in agriculture. This is a loss we can all the less afford insofar as we need an additional 28 million tons of grain output each year just to cater to the needs of population growth, let alone the demands of economic advancement and enhanced diets. While the net gain in grain output is now about 1 percent each year, population growth is nearer 2 percent.[9]

This would not matter, of course, if population growth had been restricted way back. It hasn't, and we are left with a situation with signs that we have already overshot the carrying capacity of the Earth—a critical concept to which we shall now turn.

Carrying Capacity

Carrying capacity amounts to "the number of people that the planet can support without irreversibly reducing its capacity to support people in the future."[10] While this is a global-level definition, it applies at the national level too, albeit with many qualifications as concerns international relationships of trade, investment, and so on. There is much evidence that human numbers with their consumption of resources, plus the technologies deployed to supply that consumption, are often exceeding carrying capacity already.

Consider a specific example, food production. The World Hunger Program[11] has calculated that the planetary ecosystem could, with present agrotechnologies and with equal distribution of food supplies, sustainably support no more than 5.5 billion people even if they all lived off a vegetarian diet—and the 1993 global population is already 5.5 billion. If humans derived 15 percent of their calories from meat and milk products, as do many people in South America, the total would decline to 3.7 billion. If they gained 25 percent of their calories from animal protein, as is the case with most people in North America, the Earth could support only 2.8 billion people.

True, these calculations reflect no more than today's food-production technologies. Some observers protest that such an analysis underestimates the scope for technological expertise to keep on expanding carrying capacity. We can surely hope that many advances in agrotechnologies are still available. But consider the population-versus-food record over the past four decades. From 1950 to 1984, and thanks largely to remarkable breakthroughs in Green Revolution agriculture, there was a 2.6-fold increase in world grain output. This achievement, representing an average increase of almost 3 percent per year, raised per-capita production by more than one-third: a remarkable performance by any measure. But from 1985 to 1992 there has been far less annual increase, even though the period has seen the world's farmers investing billions of dollars to increase output, supported by the incentive of rising grain prices and the restoration to production of idled U.S. cropland. Crop yields have "plateaued"; it appears that plant breeders and agronomists have (temporarily?) exhausted the scope for technological innovation. So the 1992 harvest has been little higher than that of 1984. In the meantime there were an extra 700 million people to feed. While world population has increased by almost 13 percent, grain output per person has declined by nearly 9 percent.[12]

As for the future, bear in mind that if ever there were 10 billion people to be fed adequately, we would have to produce nearly three times as many calories as today. To grow that much food, we would need to farm all the world's current croplands (remember, there is not much more agricultural territory to be opened up) as productively as Iowa's best cornfields, or three times the present world average.

All these considerations apply particularly to population growth on the part of the one billion people who live in absolute poverty. Their impoverished status and their large numbers serve to aggravate their environmental impact, and the linkages operate in multiplicative fashion, each one reinforcing the other. Moreover, these people tend to feature the highest population growth rates. Of the poorest one-fifth of developing-nation households, well over half have eight or more members, whereas at national levels the proportion is under one-third.[13] In addition, these people are unusually dependent for their survival upon the environmental base of soil, water, forests, and fisheries that make up their main stocks of economic capital. At the same time, they see scant alternative to exploiting their environments at a rate they recognize is unsustainable: they feel obliged to misuse and overuse their resource stocks today even at cost to their prospects tomorrow. They thereby undercut their principal means of livelihood—thus entrenching their poverty.[14] Population growth denies them the very inducements that could serve to reduce population growth. In turn, this appears to reinforce their motivation to have large families.[15] Thus, they face the prospect of an ever-tightening bind.

For a specific instance of carrying capacity and the impoverished, consider Nigeria. Its population of 93 million, crammed into not much more than a Texas-sized territory, lives off an average of one dollar a day. The country comprises one-fifth of all black Africans, and it is by far the biggest democracy on the continent—with all that means for security all around in a region that may soon feature one-sixth of humankind. With a population growth rate of 3 percent per year, its farming communities, which make up well over four-fifths of the population total, will have to make do in the year 2000 with less than half an acre each. The country will have to buy increasing quantities of food from outside. By far the biggest source of foreign exchange is oil. Suppose enough new oil deposits are found to maintain present levels of production and that domestic demand can be cut by one-third, but population growth persists at its present rate (it has dipped hardly at all since

1980). By the year 2015, Nigeria will need all its oil output for at-home consumption. How much economic advance and political stability will there be in a country packed with twice as many people as today—and a country where forests have almost disappeared, much soil is gone with the wind, and the Sahara is advancing?[16]

Carrying capacity can apply to economic and social factors as well, in conjunction with linkages to the environmental framework within which economies and societies operate. For a specific case, consider employment. Today the developing world's work force numbers 1.9 billion people. Of these, over one-quarter are unemployed or grossly underemployed; their total exceeds the entire work force of the developed world. By the year 2025 the developing world's work force will surely surge to well over 3 billion. To supply employment for the new worker multitudes, let alone for those without work today, means that each year during the 1990s the developing world will have to create 30 million new jobs. The United States, with an economy half again as large as the entire developing world's, sometimes has trouble in generating another 2 million jobs each year.

Again, the security connection is clear even though rarely recognized. Unemployed people lead to huge social stresses and civic tumult, especially in cities and centers of government. There are examples aplenty.

Population and Developed Nations

At the same time, let us recognize that there is a problem apart from growth of numbers in developing countries. It is growth of consumption and consumerism in developed countries. While many developed countries frequently tell developing countries to get their population act together, they do nothing about their own population growth. They do not even consider it. Not one developed country has a specific policy to indicate how many people it would like to have by such-and-such a date with such-and-such a standard of living—and hence with such-and-such an impact on the environment. Environmental impact refers especially to two of the biggest problems of all, ozone-layer depletion and global warming—impacts that reflect (in part at least) the continuing population growth of developed nations. So how about a population policy for developed countries, too?

This option is rarely considered. Indeed, it is hardly ever perceived as a worthwhile option at all. For sure, there is some ostensible justification for the lack of population policies on the part of developed nations. Their average population growth rate is 0.5 percent per year, in comparison with a developing-world average of 2.0 percent. Developed-world populations now total 1.2 billion people, or less than one-quarter of the global total. They are projected to grow by only 44 million by the year 2000, whereas developing-world populations are projected to grow by 740 million. The respective figures for 2025 are 162 million and 2.9 billion. So is the main population-growth problem not confined to developing nations? That depends on how one assesses the repercussions of population growth.

It is necessary to consider the entire impact of population growth on the planetary environmental base that ultimately sustains all societies. How about the environmental impact of Americans, who consume almost fifty times more environmentally based goods and services per capita than do the Chinese?[17] What is the comparative significance of Americans' population growth rate, 0.8 percent, and that of China, 1.3 percent? To feed one American requires three times as much agricultural produce than to feed one Chinese—with all that implies for pressures on agricultural lands. In overall consumption terms and in relation to the developing world, an American family, nominally two children, is more like forty.

Or consider it a further way. Bangladesh now has 114 million people, with an annual growth rate of 2.4 percent. The total is expected to expand in 1993 by 2.7 million people. The United States has 258 million people and a growth rate of 0.8 percent. So the American total will expand by 2.1 million. But each Bangladeshi consumes commercial energy equivalent to only three barrels of oil per year, and each American fifty-five barrels. So the population-derived increase in Bangladesh's consumption of oil equivalent will be 8.1 million barrels, that of the United States 116 million barrels. In terms of fossil energy's contribution to global warming (see Chapter 11) and hence as concerns global carrying capacity, the United States plainly bears by far the larger responsibility. It is also relevant to note that Bangladesh could lose a whole chunk of its territory to sea-level rise in the wake of global warming.

So could the time be coming when Americans should ask themselves how many people are good for America? And how many Americans are good for the world? After all, Americans are sometimes inclined to comment on developing countries' population growth in terms of its impact on the entire international order. The same question applies to Britain, Germany, Japan, and all other developed countries.

Of course there are other ways to reduce the environmental impact of developed-world citizens. They could simply consume less. Or they could utilize less polluting materials such as fuel-efficient cars—or better, leave their cars at home and use buses or get out their bicycles. Neither of these options appears to have much appeal at all. Could, then, the most likely alternative lie with reducing population growth? Ironically it would need no Draconian measures to interfere with the number of children they want to have. It would generally mean no more than eliminating all unwanted pregnancies.

This question of population growth in developed nations has not been addressed to date. Indeed it is hardly ever raised. But an implicit answer is being continuously supplied by developed-world people's use—often misuse and overuse—of their environmental base, increasingly with spillover consequences for all other countries. The problem is being resolved by default rather than by design. Since it is built into developed-world people's affluent and effluent life-styles, should these countries not move from implicit action (or inaction) to explicit response as concerns their population growth?

Security Concerns

Finally, the security connection: how does population growth tie in with the overriding issue? We cannot say there are any direct and definite linkages that make population entirely responsible for conflict. Experts argue it back and forth for hours and decades, as I have found while attending conferences on the topic over the years. It would be difficult to demonstrate an open-and-shut case in any particular instance. But it would be more than difficult to show that population growth has not been a key component of the myriad recipes that brew up strife, antagonism, confrontation, violence, and conflict—even war and revolution.

After all, if there were half as many people in India and Bangladesh, would they not suffer fewer conflicts? If there had been half as many people in El Salvador, would there have been the same population pressures that pushed multitudes into next-door Honduras? If Ethiopia's highlands had not been so overloaded, would there have been such food shortages that toppled the emperor off his throne and impelled throngs of impoverished peasants down the slopes toward Somalia? If Kenya's numbers had not expanded three times since independence, would it now be experiencing such political upheavals? If the populations of Jordan, Syria, and Iraq were not growing at a rate that would double them in twenty years or less, would they need so much water—that liquid that can ignite the flames of war like gasoline on firewood? In each case, population is not the sole factor. But whether it is merely a contributory factor or the clinch factor, it counts.

All this reminds me of another population-and-security illustration, albeit apocryphal. It concerns the time when Russia invaded China. On the first day the Russians took one million prisoners. Would the Chinese like to give in? No. On the next day the Russians took another one million prisoners. Same question, same response. Third day, same question—but a different response. "Do you Russians want to give in?"

We could discuss and debate the whole issue until the end of the book. Indeed piles of books have been written about it.[18] Frankly I grow weary of the arguments in academic citadels of Europe and the United States. If those "experts" were also experts in living in Bangladesh's overcrowded countryside, would they want to spend even an hour on still more analysis? They would prefer to be saved than studied.

11

Ozone-Layer Depletion and Global Warming

> The biosphere recognizes no division into blocs, alliances, or systems. All share the same climatic system and no one is in a position to build his own isolated and independent line of environmental defense.
>
> —Eduard Shevardnadze, 1988

Of global scope to match that of mass extinction of species but with far speedier impact is the damage we are inflicting on the world's atmosphere and climate. Probably ozone-layer depletion and certainly global warming will affect nations everywhere. Nobody on Earth will be safe from them. The ultimate impacts remain uncertain—except that they will surely be profound. If tens of millions of people die as a result of ozone-layer depletion, this will be equivalent to an extreme bombardment from the air; and if some nations lose entire portions of their territory or are even eliminated outright by sea-level rise, this will be equivalent to extreme takeover by a foreign invader. Not a shot will be fired to precipitate the two problems. But if some nations find themselves subject to such extreme injury, might they not eventually feel inclined to take aim at the principal sources of the problems?

Ozone-Layer Depletion

The ozone layer lies mainly in the stratosphere, extending from nine to thirty miles above the Earth's surface. If it were compressed, it would be a fraction of an inch thick. Yet this flimsy screen protects us from the sun's harmful ultraviolet (UV-B) radiation. A 1 percent thinning of the ozone layer leads to a 2 percent increase in the radiation dose. This will mean many adverse effects on human health, including cancers and eye cataracts.[1] There could eventually be tens of millions of extra skin cancers, many of them fatal, among people alive today.

More important, depletion of the ozone layer will depress humans' immune systems. This will leave us more susceptible to established diseases such as herpes and AIDS, and vulnerable too to the new array of diseases, tumors, and parasites we can expect in a greenhouse-affected world—tropical diseases will spread into temperate-zone communities, which will have no built-in resistance to them.

Most important of all for humans, albeit in an indirect sense, will be UV-B's impact on crop plants on land and phytoplankton-based food chains in the seas. Experiments show that enhanced UV-B cuts crop yields by anywhere from 5 percent for wheat to 90 percent for squash, with other sensitive crops in between. Soybean, for instance, a prime source of protein and our fifth most important crop worldwide, could lose one-quarter of its crop growth. Others that appear to be unduly vulnerable include barely, peas, and cauliflower.[2] As for phytoplankton, the most abundant type of ocean life in both weight and sheer numbers (one gallon of seawater can contain millions of phytoplankton plantlets), no other life form appears to be so susceptible to UV-B radiation. As phytoplankton disappear, so will zooplankton (microscopic animals that feed on them), then fish, fish-eaters, and so on to the end of the famished line.[3]

Ozone is destroyed by chlorofluorocarbons (CFCs), chemicals that serve numerous industrial purposes and are exceptionally cheap. During its lifetime aloft lasting several decades, a single CFC molecule can gobble up 100,000 ozone molecules.[4] When CFC production was at its height in 1987, the developed nations produced over 90 percent of the global total (the United States over one-third), and they consumed almost as much (the United States well over one-quarter). The developing nations accounted for the meager rest.[5] Thus the developed nations, with 22 percent of the world's population, produced and consumed CFCs at a per-capita rate 25–30 times the developing nations' average.

If recent trends of CFC production were to continue, however, the developing nations' share of CFCs production would rise to almost one-third of the global total as soon as the year 2000.[6] Much of the growing demand is centered on refrigerators, this being the largest and fastest-growing use of CFCs in developing countries. India, with 150 million households, has only 15 million refrigerators. It has been producing CFCs at a per-capita rate of only $\frac{1}{260}$th the rate of the United States.

China aims to produce even more refrigerators. To date only one Chinese household in ten possesses a refrigerator. The government plans that all its 250 million households shall have one by 2000 or shortly thereafter. To that end, China has built twelve CFC-production plants and has hoped to expand CFC production tenfold (though that would still leave per-capita output at only one-fifth that of the United States).[7] Were China to press ahead with its plans, it would become the world's largest producer and consumer of CFCs, completely offsetting the United States' efforts to phase out all its own CFC use within just a few years. The United States would have no defense against the UV-B radiation stemming from China's activities.

Fortunately the international community has taken steps, through the Montreal and Helsinki Protocols plus follow-up measures, to phase out CFC production by the year 2000 (the United States and Europe by 1995).[8] This initiative has won support from most leading CFC producers, both actual and potential. Already CFC production has plunged by half: a success story for which we can give ourselves a thumping slap on the back. Better news still, India and China have indicated their readiness to join in the great global endeavor.

But CFC substitutes will be several times as expensive as their ozone-depleting counterparts. So the developed nations have set up a fund of $240 million to help developing nations make the switch. Laudable as this gesture is, however, it is hardly more than a gesture. Future costs will eventually run many times higher. While facing the ultimate bill, we should of course bear in mind that an outlay of tens of billions of dollars over a period of decades would still be a giveaway as compared with the concealed costs of inaction. The United Nations Environment Programme estimates that had ozone-layer depletion continued uncontrolled, the price tag (through crop and fisheries losses and so on) for the United States alone could have climbed to $175 billion by 2075; and if human mortality were assessed at usual compensation rates, the costs would have been tens of times higher.[9]

Regrettably, there is evidence that the present cuts will prove to be too little and too late. The Antarctic "hole," with ozone loss for part of the year reaching half the total, has increased 13-fold over the past ten years until it now spreads across an expanse four times larger than the United States. Not only is the hole growing bigger year by year, it appears earlier and lasts longer. It already affects New Zealand and southern parts of Australia, Argentina, and Chile. Phytoplankton growth in the southern ocean is down by one-eighth, with adverse effects for zooplankton and hence for fish and other links of marine food chains; and there are reports from Chile of increased skin complaints in humans, blindness in sheep, horses, and rabbits, and deformities in tree buds.

More serious if only because it affects many more people is the hole that is emerging over parts of North America, Europe, and northern Asia with their huge populations. Northern sectors of the ozone layer are thinning twice as fast as previously assumed, and 12 percent losses have been reported—which, if they are maintained, would lead to more than 1.5 million additional cases of eye cataracts per year and a one-quarter increase in skin cancers. Worse, the depletion rate could well double again by the end of the 1990s, and even double again during the first half of the next century until the loss could eventually approach the 50 percent level found over Antarctica. The present depletion in the Northern Hemisphere lasts until well into spring, the time when people are out of doors more and when newly growing crops are at their most vulnerable.[10]

This supplies all the more reason for CFC production to be halted without any further delay whatsoever. In any case, even if all production ends as planned by 2000, CFCs already aloft or still to be injected into the atmosphere will persist long enough for ozone holes to fail to close in fewer than forty years. It will take twice as long at best for the ozone layer to be fully restored.

For a conclusion from all this, let us heed Richard Benedick, the former State Department official who led the way in putting together the original stop-CFCs package: "The concept is not obvious. A perfume spray in Paris helps to destroy an invisible gas in the stratosphere and thereby contributes to skin cancer deaths and species extinctions half a world away and several generations into the future. Neither traditional environmental law nor traditional diplomacy offer guidelines for confronting the situation. Perhaps the most extraordinary aspect of the treaty has been its imposition of substantial short-term economic costs to protect human health and the environment against unproved future dangers. At the time of the negotiations and signing, no measurable evidence of damage existed. Many governments whose cooperation was needed displayed attitudes ranging from indifference to outright hostility. International industry was strongly opposed to regulatory action. Military strength was irrelevant to the situation. Economic power also was not decisive. No single country or group of countries, however powerful, could solve the problem. Without far-ranging cooperation, the efforts of some nations to protect the ozone layer would be undermined."[11]

This statement presents a lesson for security analysts. A single large nation such as India or China can confound the best efforts of other nations to safeguard the security of their citizens. For the first time, and presumably for all time henceforth, and after decades of impotent protests at the unfairness of North–South relations, developing nations have found a way to exert hefty leverage on developed nations. For all its economic might, its technological prowess, and its military muscle, the United States, like all other once dominant powers, is powerless to defend some of its most basic interests except through holding out the hand of true cooperation with the rest of the global community—cooperation that will eventually range from CFCs to commodity prices, from ozone-layer protection to trade protection, and much more besides.

Global Warming

The second global threat from everybody's skies is even more potent and far more difficult to tackle. Buildup of so-called greenhouse gases in the atmosphere appears set to bring on a phenomenon of global warming.[12] While there is much uncertainty about just how fast it is arriving and precisely how it will affect this or that region, there is virtual certainty—according to a massive consensus of the best climatologists in the world—that it is on its way. There is also virtually across-the-board agreement that it will be bad news for all of us, whether directly or indirectly.

If we continue as we have been doing, we shall end up within a few decades with a doubling of "carbon dioxide equivalents," that is, calculating other greenhouse gases such as methane and nitrous oxide in terms of their global-warming capacity in comparison with the gas that contributes half of all global warming, carbon dioxide. This will raise average temperatures worldwide by 4.5 to 10 degrees Fahrenheit. (While that may not sound like much, let us recall that when there was a temperature shift only a little larger in the other direction a few thousand years ago, it made the difference, as Vice-President Albert Gore puts it,[13] between having a nice day and having a mile-thick layer of ice around.) Warmer climate will mean that plants will need more water to offset greater evaporation. Fortunately it looks like there will be a slight increase in rainfall worldwide, but some areas will receive rather more and others a lot less.

True, the 1992 Climate Convention looks toward ways to slow global warming. This is the international treaty that was almost watered down out of existence by President George Bush; while he talked interminably about the new world order, he was triggering a world climatic disorder of exceptional size.

All other industrialized nations are ready to go much farther than the United States along the remedial track. But the best proposals envision only a 12 or 15 percent cut in carbon dioxide emissions by 2000, whereas we need a 60 percent cut forthwith if we are to halt global warming in its tracks—and even then we would have to cope with the fair amount of greenhouse effect that is already in the pipeline as a result of past emissions of greenhouse gases.[14] Essentially, then, we are continuing with business as usual.

Which are the main sources of greenhouse gases? North America and Western Europe, while comprising only one-eighth of global population, are responsible for two-thirds of the fossil-fuel carbon dioxide emissions that account for half of global-warming processes. The United States, with only 4.7 percent of global population, accounts for 23 percent; on a per-capita basis, the American share is 5 tons per year, as against a global average of less than 1 ton. Developing nations, with 77 percent of the world's population, contribute only a small share of fossil-fuel emissions of carbon dioxide. But they contribute 30 percent of total carbon dioxide emissions taking into account tropical deforestation as well as fossil fuels. If their carbon dioxide emissions continue to grow at the rate of the past forty years, the per-capita amount will more than double during the forty-year period 1985–2025, and the overall amount will be four times larger than the developed nations' total today.[15] Just their population increase will generate another 6 billion tons per year by 2025, or three-quarters as much as the global total today. By that time, too, developing nations would be accounting for two-thirds of all emissions (which then would be much larger in total).[16]

In the meantime there is no doubt where the prime responsibility lies for global warming. The developed nations have been emitting sizable amounts of carbon dioxide for best part of one century. Were China and India, with nearly half the developing world's population, to consume fossil fuels at the per-capita rate of Western Europe (which is way below the North American rate), greenhouse gas emissions would jump by two-thirds, propelling us into instant climate disaster.[17]

Nor can there be any doubt about the seismic shift in international relations as a result of the newfound "carbon clout" of certain developing nations.[18] China, possessing one-sixth of the world's coal stocks, plans to boost its present coal consumption of 1 billion tons a year to twice as much within another quarter century. Today China contributes less than one-tenth of all greenhouse gases, of which carbon dioxide amounts to one-half. Within another quarter century, however, China could be emitting three times as much as the United States, even though its per-capita amount would still be less than that of Americans. In this sense, China can exert the capacity to wreck the climate for everyone. In today's world with its seamless atmosphere where greenhouse gases wander as they will, a nation does not have to arm itself with missiles to threaten Armageddon from the skies.

A second greenhouse gas, methane, is more potent as a greenhouse gas than carbon dioxide. Molecule for molecule, it is twenty times more effective. Worse, its accumulation in the atmosphere is increasing faster than that of carbon dioxide. But this time the main source is the developing world. Half of all humans' methane emissions come from rice paddies, among other irrigated lands, and from ruminant livestock. These two sources have been expanding largely to meet the needs of more people to feed, and to meet demand for improved diets. Regrettably methane emissions cannot be readily reduced (by contrast with the case for carbon dioxide) since they do not reflect so much wasteful consumption. Rather they are likely to keep on expanding to keep pace with population growth: scientists expect an increase of almost half in meat and dairy production by 2025, with a parallel increase in methane emissions.[19]

Global warming will exert profoundly harmful impacts on the world's, and especially developing nations', capacity to grow food. Preliminary analysis suggests it could well reduce present croplands by as much as one-third (within a range of 10–50 percent)[20]; and droughts could cause a one-tenth drop in grain harvests on average three times a decade.[21] The latest and much the most extensive assessment envisions a 10–15 percent decline in grain harvests, with substantial shortfalls for many other crops, in large parts of the tropics, plus harvest declines—possibly moderate, likely large—in North America.[22] According to some scenarios,[23] the United States' grain belt could all too easily come unbuckled; in the Southeast as much as half of the farmlands could fail; California, with the world's eighth biggest economy and heavily dependent on irrigated agriculture, could encounter huge problems in gaining enough water for crops, let alone its vast urban communities; and many of the country's woodlands and forests would likely fade from the scene.

It is in developing lands of the tropics, however, that global warming's impact could be especially severe. These regions tend to be more vulnerable to climatic change simply because they are developing.[24] At the same time, they have next to no food reserves for the most part, and their citizens often subsist off marginal diets already. Worst of all, these nations have all too limited capital and infrastructure with which to adapt to changing climate. Yet according to the latest scientific assessments,[25] the regions that appear to be at greatest risk of extreme climatic dislocations for agriculture are often those where marginal environments sometimes make agriculture an insecure enterprise already: the Sahel, southern Africa, the Indian subcontinent, eastern Brazil, and Mexico. Latest climate models show patterns of drought increasing in frequency from 5 percent of the time under the present climate to 50 percent by the year 2050.[26]

What can we do to avoid this outcome? Lots. In the United States, the single largest source of carbon dioxide whether in per-person terms or total quantity, there is much that can be done right away. Far from costing an arm and a leg, it would give a hefty leg up to the American economy. The main cost would be in people's heads rather than their pockets. Consider, for instance, the car culture, which accounts for one-quarter of carbon dioxide emissions.[27] Americans drive 2 trillion miles a year, or 8000 miles for every man, woman, and child. The typical new American car gets only 27 miles per gallon of gasoline. Less than a decade ago the figure was a little higher, but the White House has consistently allowed it to decline in response to Detroit's penchant for bigger cars. Not that bigger cars are safer. While the average weight of a new car fell by about 1000 pounds during the period 1975–88, traffic fatalities dropped by two-fifths.

New cars in Europe and Japan regularly get 50 miles per gallon, and prototypes of Volkswagen, Volvo, Toyota, and others get up to 100 miles on the open road. If Americans were to improve their car technologies to catch up with their competitors, the result would be tens of billions of dollars per year pouring into car owners' pockets from gasoline savings.[28] (There would also be savings of almost $100 billion a year from reduced disease caused by car-exhaust pollutions.) This is not likely to happen as long as oil and other fossil fuels are subsidized by the American government to the tune of $30 billion a year, or thirteen times as much as support for energy alternatives and energy-efficiency research. There is a further subsidy, an indirect one this time, in the form of $90 billion from the government for highway construction.[29] To cite Amory Lovins, the innovative analyst of the Rocky Mountain Institute in Colorado, the bottom-line answer to this situation is to get Detroit off welfare.[30]

Because of oil subsidies and ultra-low taxes, gasoline in the United States costs less in real terms today than at any time since World War II, with the biggest drop occurring since 1980. The tax component of a gallon of gasoline is only one-seventh what it is in Europe, where drivers pay anywhere from

$2.70 to $4.45 per gallon, in contrast with the American price of around $1.35.[31] By consequence, Americans are taking a free ride on everybody's climate. The correct response is to beef up the tax item, probably as part of a general carbon tax on all fossil fuels.[32] Just a $1 tax per barrel of oil and its equivalents would generate $50 billion a year worldwide—a figure to bear in mind when we consider the price of tackling global warming (see below). Europeans aim for a tax of $10 per barrel by the end of the decade.

Much the same considerations apply to building insulation, industrial motors, and other oil-thirsty sectors of the American economy. Consider home appliances, for example. Recent advances in efficiency are saving Americans $30 billion a year on their energy bills and obviating the need for extra power plants costing between $25 billion and $50 billion.[33]

In any case, Americans pay far less than they should for their energy. Hidden costs of oil, coal, natural gas, and electricity include not only subsidies and inadequate taxes but tax credits and health expenditures (to counter pollution's effects such as emphysema) as well as the biggest concealed cost of all, global warming. Not counting the last item, it is estimated that the deferred costs to society run to somewhere between $100 billion and $300 billion a year.[34] The latter figure would be equivalent to a tax of $2 per gallon of gasoline.

This is not to say that the United States has not made sizable strides in energy efficiency and conservation. For much of the time since our good friends the Arabs first taught us something about the realistic price of oil in 1973, Americans have shown a fine degree of technological ingenuity in order to wring more work from every drop of oil and chunk of coal. All in all, they have saved themselves almost $1 trillion in energy costs during the past two decades[35]; in the single year of 1992, it was a full $150 billion. If they were to rival the efforts of Europeans, however, the 1992 sum could have been $200 billion, and if they were to match the Japanese, it would have been $300 billion. To put this last figure in perspective, remember that it is equal to the Pentagon's budget and the federal deficit.

Still greater largesse lies ahead merely by making full use of existing technologies. Americans' electricity demand could be chopped by well over two-fifths before the end of the 1990s.[36] Eventually, and taking account of likely technological break-throughs, the U.S. economy could be run off well under half as much energy as today even while it expands—expands, in fact, partly because of savings through energy efficiency. This would make American products more competitive in international markets, and it would effectively put massive amounts of money into Americans' pockets.

More important still in terms of this book's theme, security (albeit of the traditional sort), note that the average American car in the late 1980s was getting only 19 miles per gallon. Had that capacity been improved by a mere 3 miles (leaving it a long way behind the average in Europe) by 1 August 1989, it would have eliminated the need for all the U.S. oil imports from both Kuwait and Iraq.[37] Plus, the costs of the war against Saddam Hussein resulted in an effective price rise of a barrel of Gulf oil from $17 to over $100. Since 1989 American oil imports have continued to climb until the country is now buying half its oil overseas, to the delight of OPEC, who believe they could hardly have wished for a more favorable trend. And the imports are equivalent to one-third of the U.S. trade deficit.

As for the other greenhouse gases, such as methane and nitrous oxide, making up one-quarter of the global warming processes, these can be partially tackled by sustainable agriculture and forestry practices that cut back on greenhouse gas emissions while also making use of soil as a sizable carbon sink rather than another carbon source. The initiative would be hardly more expensive in the long run than our present soil-mining activities.[38] As with fossil-fuel emissions, we would not come out behind, and we might find we are doing ourselves an extra good turn. So even if global warming turns out to be a fairy tale, as is still proclaimed by a few fringe fanatics, we would not lose much beyond a wasteful life-style that cannot persist for all kinds of other good reasons. Here is a situation where we would all—all nations, all generations—win together.

This is not to say that a global campaign to avoid global warming would come cheap in every respect. According to one exploratory estimate,[39] it could cost $30 billion a year to counter the more prominent causes of climate debacle. But remember: this amount is less than the United States alone spends to subsidize fossil fuels, among other forms of energy. Another estimate[40] proposes an outlay of $100 billion a year to achieve our switch not only away from carbon dioxide but from methane and other greenhouse gases. Yet that sum is only one-tenth of what we now spend to deter war, among other military activities. Wouldn't it be a solid investment to deter global insecurity of climacteric scale?

12

Mass Extinction of Species

It is not simply wrong, it is a piece of stupidity on the grandest scale for us to assume that we can simply take over the Earth as though it were part farm, part park, part zoo and domesticate it, and still survive as a species. Up until quite recently we firmly believed that we could do just this, and we regarded the prospect as man's natural destiny. We are about to learn better, and we will be lucky if we learn in time.

—Lewis Thomas

We are witnessing the start of a mass extinction of species. Earth supports at least 10 million and possibly 30 million species, conceivably many more. Already we are probably losing between 50 and 200 species per day, a rate that is 120,000 times the prehistoric rate. Worse, the rate is accelerating fast. By the year 2000, and counting from the time a few decades ago when the biological holocaust began, we will surely have lost one million species, possibly more. Worst of all, as many as half of all species face extinction by the middle of the next century—unless, of course, we do a far better job of permitting these fellow species their space on Earth.[1]

What does the massive loss of species have to do with security? Lots. As we shall see, there is a powerful economic argument for people in America to care about extinctions in Amazonia, and for Britishers to feel concern for Borneo. It affects us each dawning day as we sit down to corn flakes or munch popcorn with the evening television. Much the same when we have need to reach into the medicine cabinet. Well can scientists claim that by saving the lives of species, we are often saving our own. So there is a strong economic security for us in species: our future wellbeing is closely tied in with the wellbeing of our wild partners on Planet Earth.[2] And there are other factors that reflect security of a deeper sense still.

But first, how can we be so sure that we are losing species at such a fearful rate? There are various ways to tell. First is to look at tropical forests, which, while covering little more than one-twentieth of Earth's land surface, harbor at least half of all species. They are far and away the most species-rich biome, or ecological zone, on Earth. They are unique in another way, too. They are being destroyed faster than any other biome. When we talk about mass extinction, we are talking largely about the demise of tropical forests.

There is another way to calculate the size of the mass extinction we are imposing upon our fellow species. Six years ago I embarked upon an analysis of areas that feature exceptional concentrations of species with high levels of endemism (that is, not found elsewhere) and that face exceptional threat of habitat destruction.[3] I worked out that there are fourteen of these "hot-spot" areas in tropical forests and four in what are known as Mediterranean-type zones such as California. Altogether they contain 50,000, or 20 percent, of Earth's plant species, and a still larger proportion of animal species, in just 0.5 percent of

Earth's land surface. If, as is anticipated, they undergo wide-spread environmental destruction within the foreseeable future, these areas alone will witness a mass extinction greater than any since the great dying of the dinosaurs 65 million years ago.

All this is very regrettable. But what, the reader may still ask, does it have to do with security? To put the issue at a basic level, that of economics today, consider the many contributions of species to our material welfare.

Our Economic Stake in Species

Little though we may be aware of it, species and their germ-plasm resources support us at dozens of points in our daily rounds.

Consider America's corn. The greatest corn-growing country in the world, the United States supplies not only its own citizens with a staple food but helps to feed hundreds of millions of other people around the world. In 1970, a severe blight spread so swiftly that some states lost half their crop. Costs to corn growers, and hence to consumers, totaled $2 billion.[4] The situation was saved with new strains of corn that resisted the blight, the genetic materials coming originally from corn's native home, Mexico. In 1978, a wild relative of corn was found in a forest of central Mexico. The weedy-looking plant was surviving in just a last few acres of forest, in a montane area that was being rapidly settled by landless farmers. So the whole species comprised no more than a few thousand final stalks in all. It now turns out to offer genetic resistance to at least seven major diseases of commercial corn. Still more important, the wild plant, growing in a cool, damp environment in its montane forests, opens up the prospect of growing maize in localities that have hitherto been too cold and wet for established varieties, and hence could expand the range of corn worldwide by at least one-tenth. Commercial benefits to corn consumers worldwide could eventually be reckoned in billions of dollars a year[5]—and all this from a single forest plant that was all but extinct when it first became known to science.

Other tropical forest plants supply germplasm for genetic improvement of several other major crops such as rice, coffee, cocoa, and bananas. A while ago the famous Green Revolution rice in eastern Asia was hit by disease. Rice geneticists searched through their gene-bank collections of rice varieties, trying to find a strain resistant to the disease. No success. So they returned to wild relatives of rice, that is, those plants that are close associates of cultivated rice. After much searching, they luckily came across a variety that supplied the answer—and ironically, in a forest tract at Silent Valley in India that was shortly to be submerged by a mammoth reservoir.

So much for a couple of examples of improved forms of existing foods. There are many entirely new foods, too, waiting for us out there in the wild. Ten years ago the kiwi fruit was virtually unknown outside its tropical forest home of southern China, but now it graces the shelves of supermarkets around the world. In the forests of New Guinea there are 200 wild fruits that have proven palatable to local people, that are full of vitamins, and that could flourish in plantations. Also in New Guinea is the native home of the winged bean, a vegetable packed with more protein than the soybean and able to grow in many areas where soybeans cannot. It was unknown to the outside world until the mid-1970s, when the U.S. National Academy of Sciences highlighted its potential as a domestic crop. Today it is grown in fifty countries of the tropics and helps to upgrade the nutritional plane of people in the hundreds of millions.

There are lots of other fruits and vegetables awaiting our attention. We are losing them and other plant species in the hundreds if not thousands every year.

Much the same applies to plants' contributions to our health. Roughly half of all medicines, drugs, and pharmaceuticals purchased by American consumers owe their manufacture, in one way or another, to genetic resources of wild plants, mostly from tropical developing nations. The commercial sales of these products now amount to more than $40 billion a year worldwide. The rosy periwinkle, originating in Madagascar's tropical forests, has produced two potent drugs for use against Hodgkin's disease, childhood leukemia, and other blood cancers. Commercial sales of these two drugs now run to around $200 million a year in the United States alone, while economic benefits to American society can be reckoned as twice that much. Cancer researchers believe there are at least another twenty plant species in tropical forests alone that could be used to generate similar superstar drugs against cancer. Equally to the point, three plants offer experimental potential against AIDS. Yet as with the potential of wild foods, these economic capacities of species depend upon the scientist getting to the species before the chainsaw gets to their habitats.

Still other examples of species-derived benefits can be cited for industry and energy. Specialized plant materials contribute to industry by way of gums and exudates, essential oils and ethereal oils, resins and oleoresins, dyes, tannins, vegetable fats and waxes, insecticides, and multitudes of other biodynamic compounds. Many wild plants bear oil-rich seeds with potential for the manufacture of fibers, detergents, starch, and general edibles—even for an improved form of golf ball.

To date, however, scientists have intensively investigated only one plant species in one hundred, and a far smaller share of animal species, to assess their economic applications. We can realistically assume that there are whole cornucopias of improved and new foods, entire pharmacopoeias of new medicines, and entire stocks of industrial raw materials awaiting us in the wild, provided the species in question are not pushed over the edge into extinction.

The great bulk of extinctions are occurring in the tropics, this being where most species exist and where habitat destruction is proceeding fastest. The tropics are roughly the same as the developing world. Remote as the tropics may seem for developed nations of the temperate zones, developed nations have a pronounced stake in the issue. The United States, for instance, being "gene poor," depends heavily upon species' genetic resources from the tropics. The U.S. Department of Agriculture estimates that germplasm contributions lead to increases in productivity that average around 1 percent annually, with a farm-gate value that now approaches $2 billion. At the same time, the United States possesses hefty technological capacity to exploit genetic resources for economic benefit, especially through the emergent industry of bioengineering.

To this extent, we all have a solid security interest here. Entire sectors of our economies depend upon our fellow species. The full future of American agriculture is critically linked to dependable supplies of germplasm from foreign lands. Yet the connection is rarely recognized. Which Pentagon official has ever declaimed about the U.S. stake in species way beyond America's borders? Which may count more in the future, the Fort Bragg military base or the Fort Knox gene bank with its grossly inadequate stocks of crop germplasm?

Alas, there is vast scope here for antagonism and confrontation among the community of nations. At the Rio U.N. Conference on Environment and Development in June 1992, most nations of the world signed a Biodiversity Convention, the first time a global treaty of that sort had been placed on the table. The Convention aims to enable the world's nations to make common cause as concerns the world's biodiversity. Regrettably the developing nations, led by Brazil, China, and India, stated that since most of the planet's genetic materials live in their lands, they should be accorded preferential access to the technologies deployed for exploitation of the resources. These technologies, making up a billion-dollar industry, are mostly to be found in developed nations, which almost all rejected the developing nations' request (the resistance was led by the United States, the only country in the world to refuse to sign the Convention). Here was a chance for both sides to come out ahead if there had been more flexibility and trust. As it turned out, both sides lost together, and the Convention remains little more than a pious statement of intent. Nor does there seem any prospect that the nations of the Earth will rise above their divisions any time soon.

The Future of Evolution

Finally, let us look at another angle of species and security. It concerns the impact of mass extinction on the entire future course of evolution, no less. Sheer loss of species will be far from the whole story. It will likely turn out in the long run to be much less significant than our reduction of evolution's capacity to generate replacement species. This ultimate upshot will persist, so far as we can gather from recovery periods following mass extinctions in the prehistoric past, for 5 million years at least, probably several times longer. It behooves us, then, to consider the ultimate repercussions of the present mass extinction on the future course of evolution, no less. We appear to be giving it hardly a thought. Who ever hears a word about this aspect of mass extinction?

To put the issue in perspective, consider that Earth is subject to many other environmental assaults such as acid rain, soil erosion, decline of forests, and spread of deserts. Fortunately, all these problems are inherently reversible, and they can be rectified within a matter of decades or centuries. Even ozone-layer depletion and global warming could eventually be corrected if we were to spend enough money and time; a few centuries at most would do the job. Species extinction, by contrast, is final—for the lengthy time being at least. True, evolution will eventually come up with replacement species. But for evolution to generate an array of species that will match today's in numbers and diversity will require a recovery period extending for several million years at best.

From what we can discern from the geologic record, the "bounce-back" time following a mass extinction generally requires several million years. After the dinosaur extinction 65 million years ago, 5 million to 10 million years elapsed before there were bats in the skies and whales in the seas. In the wake of the species crash during the late Permian period 245 million years ago, when marine invertebrates, being the most numerous categories of species, lost about half their families, it took 20 million years before the survivors could establish even half as many families as were lost.

But the evolutionary outcome this time around could prove yet more drastic. The critical factor lies with the likely loss of strategic environments. Not only do we appear set to lose most if not virtually all tropical forests, but there is a steady depletion of tropical coral reefs, wetlands, estuaries, and other ecological zones with exceptional abundance and diversity of species and with unusual complexity of ecological workings. These environments have served in the past as preeminent "powerhouses" of evolution, meaning they have produced more species than other environments. Virtually every major group of vertebrates and many other large categories of animals, also plants, appear to have originated in spacious zones with warm, equable climates, especially their forests. It is likewise supposed that the rate of evolutionary diversification—whether through proliferation of species or through emergence of major new adaptations—has been greatest in the tropics, especially in tropical forests. In addition, tropical species seem to persist for only brief periods of geological time, which implies a high rate of evolution.

Furthermore, the species fallout will surely apply across most if not all major categories of species. This is axiomatic as extensive environments are eliminated wholesale. So the result will contrast sharply with the demise of the dinosaurs and associated species, when not only placental mammals survived (leading to the adaptive radiation of mammals, eventually including the human species), but also birds, amphibians, and crocodiles, among other non-dinosaurian reptiles.

On top of all this, there is a more particular question. We are almost certainly determining that there shall be no new forms of the great cats and apes, rhinos and pandas, and many other species known among biologists as charismatic mega-vertebrates. To produce offspring species, they would have to maintain population totals, and hence gene pools for natural selection to work on, many times larger than today's remnant numbers. Zoos are a limited answer. Though they increasingly do a fine job for threatened species, they will be little better than geriatric wards.

The impending upheaval in evolution's course could rank as one of the greatest biological revolutions of all time. In scale and significance, it could match the development of aerobic respiration, the emergence of flowering plants, and the arrival of limbed animals. But whereas these three departures in life's history rank as advances, the prospective depletion of many evolutionary capacities will rank as a distinctive setback.

These, then, are some ultimate issues for us to bear in mind as we begin to impose a fundamental shift on evolution's course. The biggest factor by far is that as we proceed on our impoverishing way, we scarcely pause to consider what we are doing. We are "deciding" without even the most superficial reflection—deciding all too unwittingly, but effectively and increasingly.

All this should give us pause. Within the space of the lifetimes of most readers of this book, just a few human generations, we shall—in the absence of greatly expanded conservation efforts—impoverish the biosphere to an extent that will persist for at least 200,000 human generations, or twenty times longer than the period since humans emerged as a species.

So what has this final factor to do with security? To some people, it matters far more than anti-cancer drugs or even better foods to feed the billions. It relates to our proper position on this Earth, which supports multiple life forms besides our own. It relates to our relationships with all other life that surrounds us and ultimately supports us. It tells us that we deploy an altogether too limited sense of our true security, the security that dwells within us. It is the deepest security, even if among the most disregarded.

Those who don't understand could have it explained to them fifty times and they still would not hear the message.

13

Environmental Refugees

The gravest effects of climate change may be those on human migration as millions are uprooted by shoreline erosion, coastal flooding, and agricultural disruption.

—Intergovernmental Panel
on Climate Change, 1990

During the past few years our television screens have regularly featured refugees—political and economic refugees and, increasingly, environmental refugees. The United States has had some direct and vivid experience of environmental refugees. The Haitian boat people, abandoning their homelands in part because their country has become an environmental basket case, have generally been granted entry with much of the spirit that Uncle Sam traditionally extends to the destitute—and the cost of taking in these Haitians has been more than the federal government has been spending on environmental aid for the entire Caribbean.

We are going to hear much more about environmental refugees, being people who can no longer gain a secure livelihood in their homelands because of drought, soil erosion, desertification, or other environmental problems. In their desperation they feel they have no alternative but to seek sanctuary elsewhere, however hazardous the attempt might be.[1] According to latest estimates,[2] there are at least 10 million of these destitutes today, or almost as many as all other refugees (political, religious, ethnic) combined. The figure is certainly on the low side, since governments generally take little official account of this unconventional category of refugees. Not all of them have fled their countries; many of them are "internally displaced." But all have abandoned their homelands on a permanent basis, having no hope of a foreseeable return.

To put the present figure in proportion, recall that at the end of World War II there were 7 million homeless people wandering around Europe. But while the present human tragedy is unprecedented in the wretched history of refugees, it is nothing as compared with what lies ahead in a greenhouse-affected world (Table 13.1).

Bangladesh

Consider, for instance, Bangladesh. In April 1991 we watched the ravages wrought by a cyclone on the country's coast, with the loss of 200,000 lives (possibly many more) and millions left homeless—all in one portion of a single small country. In decades ahead this will likely seem a trifling affair as one climate catastrophe after another strikes this hapless land. Bangladesh is geographically the size of Florida or England, with 114 million people today and the highest human density on

TABLE 13.1
ENVIRONMENTAL REFUGEES IN
A GREENHOUSE-AFFECTED WORLD

Country or Region	Total of Refugees Foreseen (millions)
Bangladesh	15
Egypt	15
Deltas and other coastal zones	70
Agriculturally dislocated areas	50
Total	150

Earth. The Netherlands, the most crowded country in the developed world, has a density only half as much. And the great majority of Bangladeshis live in rural areas, meaning the whole national territory is congested with people, in contrast with the Netherlands, where only one citizen in ten lives in the countryside. Because of the population factor alone, Bangladesh is exceptionally vulnerable to freak phenomena such as cyclones and coastal flooding.

Unfortunately, Bangladesh is highly susceptible to flooding by reason of its low-lying territory. More than half the country lies less than fifteen feet above sea level, leaving it more vulnerable than any other country on Earth to sea-level rise. In addition, Bangladesh straddles the floodplains confluence of three great rivers, the Ganges, Brahmaputra, and Meghna. In effect, most of the country is one great delta, where even a slight increase in water level leads to widespread flooding.

Now consider a scenario for sea-level rise in Bangladesh, worked out by Dr. Jim Broadus and his colleagues of the

Woods Hole Oceanographic Institute in Massachusetts, and confirmed by a former minister of agriculture in Bangladesh, Dr. Fasih Mahtab.[3] While global warming is projected to cause the sea level to rise everywhere around the world, it is expected to do so more along Bangladesh's coastline due to land subsidence. There could well be an effective sea-level rise of three feet, or more than twice the global average, within another few decades. In conjunction with tidal waves and storm surges, this will generate severely adverse impacts for all Bangladesh's lands up to fifteen feet above sea level, or more than half the country. Moreover, global warming is expected to cause cyclones to become stronger and more frequent,[4] producing outsize variations of tidal waves and storm surges.

Note, too, that by 2050, or roughly the time when these threats are likely to come to pass, Bangladesh is projected to have around 220 million people, or nearly twice as many as today.

The outcome is expected to be catastrophic for a country that, with a per-capita GNP today of only $200, is among the poorest in the world. That is to say, Bangladesh can deploy little in the way of engineering capacity to resist natural disasters. An entire coastal sector of Bangladesh could permanently disappear beneath the waves, and a still larger area could be regularly overtaken by 18-foot storm surges reaching 100 miles or more inland, that is, one-third of the way from the coast to the northern border. Recall that in April 1991 it was not so much the cyclone that killed 200,000 people, but rather the inrushing of seawater that drowned them. A similar disaster occurred in 1974, when 300,000 people perished—and the resultant disorders lead to the overthrow of the country's founder and president, Sheik Mujibur Rahman.

Taken together, the expected hazards could eventually destroy the homes and holdings of at least 15 million Bangladeshis.[5] Even if the catastrophe could be partially contained through engineering works of heroic scale, there would be further problems from added effects through acute congestion in the new coastal zones. Moreover, the victims could look for little help from an already poor nation that would have lost a large chunk of its economic base.

On top of all these problems stemming from the encroaching sea, there could be trouble for Bangladesh from the opposite side of the country, the Himalayas. The three great rivers pouring into Bangladesh carry super-swollen waters after the monsoon strikes the Himalayas. Their combined outflow is two and a half times that of the Mississippi, and surpassed only by that of the Amazon. (Less than one-tenth of their catchments lies within Bangladesh, leaving the country with no control over the vast volumes of water pouring into its territory.) In September 1988, river flooding left three-quarters of the country inundated, with $1.5 billion worth of damage (worth an entire year's GNP growth); the disaster had nothing to do with waters from the sea. As global warming gets under way, there will be increased differences in temperature between sea and land, causing the monsoon system to become more powerful and violent in its impact,[6] whereupon the rivers' flow into Bangladesh could expand by at least half during the monsoon season. Indeed, some experts believe the threat from inland flooding may turn out to be more serious than the better recognized threat from rising sea level.[7]

All in all, the prospect for Bangladesh looks likely to be a disaster beyond all but the worst deprivations suffered by any nation through defeat in war.

Egypt

Consider next the global-warming prospect for Egypt. The Nile River supports a six-mile-wide strip of farmlands plus the delta plain, totaling 14,000 square miles, or a mere 3.5 percent of national territory. In this area live 57 million people today, making for a local population density of 3600 per square mile, or nearly twice the nationwide density of Bangladesh. As we have seen in Chapter 3, Egypt already has severe difficulty in feeding itself, importing half of its food. But in a greenhouse-affected world, it is anticipated that drier conditions will cause a one-fifth drop in the corn yield and a one-third drop in the wheat yield[8]—a prospect that will be compounded by a huge decline in the Nile River flow.[9] Perhaps even more important, global warming will bring on sea-level rise and permanent flooding of prime farmlands. Moreover, in the worst affected area, the Nile delta, there is rapid subsidence too, expected to produce an effective sea-level rise of three feet by 2050 or thereabouts.[10]

A sea-level rise of this scale would permanently flood the delta plain for almost twenty miles inland.[11] Egypt would lose one-seventh of its habitable land and much of its main bread-basket. Already 10 million people live in that sector of the delta, which is only three feet above high tide. Given that Egypt's present population is projected to increase to 116 million by the year 2050, it is realistic to anticipate that the sea-level rise could eventually displace as many as 15 million people.[12] This prognosis, moreover, is cautious and conservative. There will be additional problems such as intrusion of saltwater up the foreshortened Nile, which will further reduce the irrigated lands that support virtually the whole of Egypt's agriculture.

Other Deltas and Coastal Zones

Many other deltas, together with estuaries and extensive coastal zones, will be sorely affected, since they are vulnerable to even a moderate degree of sea-level rise. Among such "severe-risk" areas are the mouths of the Hwang Ho and Yangtze rivers in

China, the Mekong in Vietnam, the Chao Phraya in Thailand, the Salween and Irrawaddy in Myanmar, the Indus in Pakistan, the Tigris and Euphrates in Iraq, the Zambezi in Mozambique, the Niger in Nigeria, the Gambia in Gambia, the Senegal in Senegal, the Magdalena in Colombia, the Orinoco in Venezuela, the Courantyne and Mazuruni in British Guiana, the Amazon and São Francisco in Brazil, and La Plata in Argentina, plus less extensive areas at the mouths of other major rivers.[13] Given present populations and their growth rates, the United Nations Environment Programme anticipates that at least 100 million (perhaps tens of millions more) people will find themselves flooded out, or suffering related troubles such as storm surges from time to time.[14] Yet these low-lying areas are precisely the localities that feature some of the densest human settlements and the most intensive agriculture on Earth. In the Mekong delta, for instance, 10 million people now live in areas no more than three feet above high tide.[15]

Still other large communities could be threatened in low-lying coastal territories, notably the mega-metropolises of Shanghai, Manila, Jakarta, Bangkok, Calcutta, Madras, Bombay, Karachi, Lagos, Rio de Janeiro, and Buenos Aires.[16] Their collective populations totaled 94 million in 1985, and are projected to reach 144 million as soon as 2000, perhaps rising to 200 million or more by 2050. Some of these urban areas are subject to subsidence, too; Bangkok is sinking by several inches a year,[17] a rate that if continued will amount to several feet by 2050. In fact, subsidence may well increase in many other areas as groundwater stocks are increasingly exploited for agriculture, industry, and domestic needs. For a good many of these outsize cities, subsidence is expected to contribute to a relative sea-level rise of around three feet.[18]

If only a small proportion of the projected populations in these urban conglomerations become displaced by sea-level rise and related troubles such as tidal waves and storm surges, this could add, say, 50 million to the refugee numbers.[19] When we consider as well the delta and estuary areas listed above, and factor in their present populations and growth rates, we find that at least 70 million (possibly tens of millions more) people could be periodically flooded out.[20] The total could be a good deal higher if we include smaller cities as well, since already two out of three cities with 2.5 million people are located on coasts.[21]

At the same time, a number of developed-world cities will be threatened, notably New York, Boston, Miami, New Orleans, Rotterdam, Venice, St. Petersburg, Sydney, Melbourne, Brisbane, and Tokyo. But developed nations will be able to deploy the engineering skills and the finance to hold back the sea by dykes, after the manner of the Netherlands; or they can go some way to moving their cities inland over a period of decades. (It will be expensive; the Netherlands estimates it will have to spend $10 billion, equivalent to 4.3 percent of the nation's GNP today, to hold back the sea, while the United States could well face costs of hundreds of billions of dollars to safeguard just its East Coast cities.[22]) These options will hardly be open to developing nations, simply because they are developing.

In yet broader terms, the United Nations Environment Programme anticipates that sea-level rise plus tidal waves and storm surges could eventually threaten 5 million square kilometers of coastal lands.[23] While this aggregate area is equivalent to the United States west of the Mississippi, it amounts to only 3 percent of Earth's land surface. Yet it is home to well over 1 billion people already, a total projected to rise to at least 2 billion within another three to four decades (it also encompasses one-third of global croplands).[24] As early as the year 2000, three-quarters of humankind could be living within 60 miles of coastlines.[25]

The principal countries at greatest risk by virtue of coastal plains with large human populations are Indonesia, Thailand, Bangladesh, Pakistan, Mozambique, Egypt, and Senegal.[26] A country does not have to feature a broad and shallow coastal plain to qualify for the list. Indonesia's sea shore generally gives way quickly to uplands, yet its 13,000 islands comprise coastlines totaling 51,000 miles, or four times longer than the coastlines of the United States.[27]

Not on the "leading-country" list is China, since only a smallish proportion of its population lives in the coastal zone. But because of the sheer size of China's population overall, the number in question is impressive, 76 million people today living on 50,000 square miles of lands only three feet or less above sea level (an area equivalent to New York State).[28] The government estimates that a sea-level rise of one and a half feet would eliminate the homes of 30 million people today, and that as many as 100 million people will ultimately be affected to some degree by coastal flooding.[29] Nor will there be much room for them to retreat *en masse* inland. In China's coastal plain live 350 million people today, with a congestion quotient as high as any on Earth.

Similar considerations apply to India, with a coastal-zone population today of 180 million.[30]

The aggregate area cited, almost 2 million square miles of coastal lands, includes those sectors of Bangladesh and Egypt already considered above. So let's suppose that other areas would supply many millions of refugees. Plainly this estimate is rough indeed; it is advanced with the sole purpose of enabling us to get to preliminary grips with an issue of exceptional importance. It is equally plain that this figure is conservative: eastern China and India could readily generate tens of millions each.

Let us postulate as an overall reckoning that all these other deltas and coastal zones generate a further 70 million refugees. Note too that the calculation is cautious and conservative throughout. Of course, this is rudimentary arithmetic. But however wide of the mark it may be in the practical outcome, it surely serves as a working figure for the present purpose—

which is, and to emphasize the main point of this analysis, an attempt to assess the ultimate size of what is potentially one of the biggest tragedies to overtake humankind.

Island States

Also at risk are a number of island states such as the Maldives, Kiribati, Tuvalu, and the Marshalls in the Indian and Pacific oceans, plus a dozen or more such states in the Caribbean. The first group will be totally vulnerable to sea-level rise and flooding insofar as virtually their entire territories are only a few feet above sea level. They face the prospect of outright elimination: they will simply cease to exist as nations, though their combined populations will hardly total one million people.[31] The other islands, notably in the Caribbean, may become subject to tropical storms of greater frequency and intensity, sufficient to render parts of the islands much less suitable for permanent human habitation.

Agricultural Dislocations

On top of all this there is the prospect of other greenhouse effects, such as shifts in monsoon systems and the arrival of continuous droughts—with all that both entail for our capacity to feed ourselves. A temperature rise of only 1.8 degrees Fahrenheit, entirely likely by early next century, could affect monsoon patterns to an extent that would dwarf the direct drought effects of such a temperature rise.[32] The area most vulnerable to monsoon dislocations is the Indian subcontinent, projected to feature 1.4 billion people by the year 2030. India depends upon the monsoon for 70 percent of its rainfall. In broader terms, the entire Asia–Pacific region is unduly vulnerable to monsoonal changes if only because it contains well over half the world's population today, projected to become a still larger proportion by 2030.[33]

As for drought and its repercussions for agriculture, this is more of an "iffy" business, since the climatic quirks are less well predicted at a regional level than is the case with monsoon patterns or sea-level rise. Areas considered to be susceptible include much of the United States, southern Canada, northern Mexico, northern Chile, northeastern Brazil, eastern Argentina, southern Europe and the rest of the Mediterranean basin, the Sahel, the southern quarter of Africa, sectors of middle and tropical latitudes of Asia, and Australia.

Note the results of an exploratory analysis of the drought prospect by Paul Ehrlich and Gretchen Daily at Stanford University.[34] They conclude it is entirely plausible that early next century there could be a 10 percent reduction in the global grain harvest three times a decade. The 1988 droughts in just the United States, Canada, and China resulted in almost a 5 percent decline (in America's grain belt, it was 30 percent). A mere 1.5 degree Fahrenheit increase in temperature could reduce India's wheat crop by 10 percent.[35]

Given the way the world's food reserves have dwindled almost to nothing as a result of the droughts in the 1980s,[36] it is not unrealistic to reckon with the Ehrlich team that each such grain-harvest shortfall would result in huge numbers of starvation deaths—anywhere, according to the computer-model calculations, from 50 million to 400 million people. Mega-scale famines are held at bay today in part through food shipments from the great grain belt of North America, serving to close the food-population gap in more than a hundred countries. In a greenhouse-affected world the grain belt would surely come unbuckled to an extent that there would likely be fewer such shipments as Americans find it harder to feed themselves, let alone the rest of the world.

This analysis has been reinforced by a still more recent and much more detailed assessment,[37] which postulates that global warming will eventually reduce grain production, especially in the tropics (meaning, mainly the developing world), by 10 to 15 percent. In conjunction with other factors such as population growth and increased grain prices to reflect scarcity, there will be a projected expansion of 640 million in the number of hungry people, bringing the famished total to well over 1 billion.

Starvation crises of altogether unprecedented scale will trigger mass migrations of people from famine-stricken areas. How many is impossible to say with even a modicum of precision. But for the sake of getting a handle on what could become one of the most significant phenomena of the coming decades, it is surely reasonable to hazard an informed estimate—nothing more, nothing less—of 50 million refugees. Possibly, or probably, many more. There could be 50 million in Africa alone.[38]

Aggregate Assessment

These, then, are some estimates of the numbers of environmental refugees that could well become an everyday part of our future world. Of course they are very much a case of best-judgment estimates—exploratory at most. Some of them could be off target by tens of millions either way. But the combined total of future refugees is 150 million. However rough the reckoning, it supplies an initial insight into the size of the upcoming problem of environmental refugees. The total would amount to almost one person in fifty of the 10 billion people expected to be on Earth by the time the greenhouse effect is exerting its worldwide impact.

The consequences would be among the most important of all in a greenhouse-affected world. We could eventually (or soon?) witness multitudes of despair-driven migrants heading from tropical Asia toward "empty" Australia, from China toward

Siberia, from Latin America toward North America, and from Africa toward Europe. Indeed, a mini-migration has already begun: almost one million people make their illegal way across the United States' southern border each year. Other nations are likely to be little more able than the United States to control their borders. Bear in mind, too, that the refugees would feel justified in seeking sanctuary in developed nations on the grounds that it would be the developed nations that would have largely set up the problem of global warming.

The repercussions would be profound. Refugees often arrive with what is perceived by host communities as "unwanted luggage" in the form of alien customs, religious practices, and dietary habits, plus new pathogens and susceptibility to local pathogens. Resettlement is generally difficult, full assimilation is rare. Economic and social dislocations would proliferate, cultural and ethnic problems would multiply, and the political fallout would be extensive if not explosive. We are familiar enough with the strains generated for receiver countries when they have to face throngs of refugees fleeing from drought, famine, floods, and other disasters. To cite a former United Nations High Commissioner for Refugees, Prince Saddrudin Aga Khan, refugees form "a perfect recipe for widespread human suffering, social disorder, and political instability."

In mid-1992 Pakistan was struggling to play host to 3.3 million refugees (political and religious as well as environmental refugees). Ethiopia was trying to do the same for 710,000 refugees; Sudan, 678,000; Mexico, 356,000; and Costa Rica, 278,000. All of these countries had more than enough on their plates to support their own citizens. The United States and Canada found themselves accommodating 1.5 million refugees and Europe almost 900,000.[39] It now costs developed nations $8 billion a year to cater to refugees, or a full one-sixth of all foreign aid they supply to developing nations. It would be a handsome payoff investment to boost foreign aid and tackle much of the refugee problem at the source rather than wait and pick up a far heftier tab through responding to symptoms of the same problem.

Yet our experience to date will prove a pale portent of what lies ahead. Thus far refugees have been viewed as a peripheral concern, a kind of aberration from the normal way of things. In the world of the future, they are likely to become a predominant feature of our One Earth landscape. It requires a leap of the imagination to envision as many as 150 million refugees.

Yet amid the din of international debate on global warming, we hear next to nothing about environmental refugees. Few people, let alone political leaders, think much about them. One exception is the Home Affairs minister in Italy, who invited me to Rome to discuss the prospect—on the grounds that Italy would be in the front line of a refugee invasion across the "Mediterranean pond." He did well to consider the prospect. The countries of the Mediterranean basin today total 400 million people, roughly 200 million on the European side and another 200 million in the Middle East and North African countries, mostly countries with fragile economies. By the year 2030 the European population is expected to remain more or less the same, but the others are projected to almost double. By 2030, moreover, it is all too likely that global warming will have extended the Sahara northward until it overtakes the North African shoreline; and the eastern Mediterranean countries may likewise be even more arid than today. There is good reason to suppose that multitudes of impoverished people will be trying to make their way into Europe. Already there are 15 million legal migrants in the European Community alone, and at least 5 million illegals, almost all from Mediterranean countries. In Germany one in four social-security claimants is a non-German, and asylum seekers cost the state almost $3 billion a year.

Environmental refugees totaling anywhere between 100 and 200 million would prove a deeply destabilizing factor in international relations. This would be all the more the case if the world were battling to cope with a plethora of other environmental problems at the same time. Whatever the justice of their cause and the urgency of their plight, recent experience with a mere 10 million such refugees shows they would scarcely be welcome in whatever place they seek sanctuary. As a measure of the political upheavals they engender, note that there are now more Mauritanians living in Senegal and Mali than in Mauritania itself, a situation that grossly aggravates the political stresses in a volatile sector of northwestern Africa.

In short, environmental refugees would be seen as an unacceptable burden on host countries. Governments would try to shuffle off responsibility, and cries of blame would echo back and forth among the global community. All too easily the situation could become intolerable—on both sides. Multitudes of refugees would undermine security in a manner to equal any but the greatest threats we consider in this book. And the sheer tragedy in human terms would surpass anything we have known in the history of humankind.

14

The Synergistic Connection

Tolerance of one environmental stress tends to be lower when other stresses are working at the same time. A plant, for instance, subjected to reduced sunlight and hence less photosynthesis, becomes unusually prone to the adverse effects of cold weather, and suffers much more in the cold than would a plant enjoying normal growth and full vigor. A similar expanded effect works the other way round as well. Much the same applies to many other ecological linkages with their compounding impacts. So powerful can these impacts be that the overall result is often ten times greater than the mere sum of the parts.

—O. L. Lange,
Encyclopedia of Plant Physiology (1981)

What are synergisms, and what have they ever done for us? Few environmental scientists could tell you much about them, since they remain something of a black hole of research. But we may soon come to hear a good deal about this phenomenon with its ungainly name and its emphatic impacts.

Probably the biggest environmental problem of all on the horizon will turn out to be that of the interactions between lesser problems. The phenomenon arises when two or more environmental processes interact in such a manner that the joint product of their interactions is not merely additive but multiplicative; that is, their impacts operate in a mutually amplifying fashion. One problem combines with another, and the upshot is not a double problem but a super-problem.[1]

Yet despite their obvious importance, we know all too little about environmental synergisms. Ecologists cannot even identify and define their main manifestations in nature, let alone document their more frequent repercussions. To the extent that we can discern some of the possible synergistic mechanisms at work in the environmental upheavals ahead, the better we shall start to understand some potential reactions— and the better we shall be able to anticipate and even prevent some of their more adverse repercussions.

What synergisms may arise among environmental problems in the foreseeable future? Consider the question of ozone-layer depletion and global warming. There is a well-known connection between these two phenomena, namely, the role played in each by chlorinated fluorocarbons. Not nearly so recognized is the grossly aggravating impact of the first on the second. Ozone-layer depletion leads to increased ultraviolet radiation (UV-B), and no group of organisms is more susceptible to UV-B than marine phytoplankton. It is these phytoplankton that serve as one of the main mechanisms for the oceans to take up carbon dioxide from the atmosphere. The oceans are considered to absorb roughly half of all carbon dioxide emitted into the atmosphere each year. Were marine phytoplankton to be markedly reduced through an increase in UV-B, the oceans' capacity to serve as the major sink for carbon dioxide would in turn suffer significant decline. In turn again, this would generate a sizable acceleration in greenhouse-effect processes. The UV-B impact on the phytoplankton would not be a fraction so significant for the global environment if it were not for the carbon-absorbing role of the phytoplankton.[2] Alternatively stated, the first factor would not be nearly so influential (no more than additively influential) without the second factor that predisposes the first to an importance it would not otherwise possess.

Much the same applies to another great carbon sink, the trees of the forests that extend right around the Northern Hemisphere. There is preliminary evidence that certain tree species could by damaged by UV-B, leaving them less capable of photosynthesis and hence less able to soak up carbon dioxide.

A similar synergism will operate between global warming and decline of genetic variability for our crop plants. In a greenhouse-affected world, many agricultural regions look likely to experience sharply changed environmental conditions, notably higher temperatures and reduced soil moisture. Yet most of our agricultural crops are finely tuned to present climatic conditions. Hence there will be a need to expand the genetic underpinnings of our crops—and this will place a premium on germplasm variability to build up, for example, drought resistance. The same applies to genetic adaptations for crop plants to counter new pests and diseases, such as are likely to thrive in a greenhouse-affected world.

Yet we are permitting the gene reservoirs for crop plants to be depleted at unprecedentedly rapid rates, to an extent that already leaves our crops dependent upon a critically reduced genetic-resource base. To emphasize the synergistic dimension: while genetic erosion is already serious, it will become much more serious when combined with global warming. Conversely, too, greenhouse-affected crops will be in much deeper trouble if their adaptability cannot be reinforced with a proper range of genetic variability.

For another example, consider acid rain and tropical forests. Acid rain is already a recognizable problem in the moist forests of southern China, and will likely soon affect other tropical forests in Sumatra, Peninsular Malaysia, south-central Thailand, southwestern India, parts of West Africa, northern Venezuela, and southeastern Brazil.[3] Still other tropical-forest countries want to embark on rapid industrialization, offering scope for a still wider spread of acid rain. Acid rain takes no notice of human dispensations such as parks and reserves; it will make it vastly more difficult to protect tropical forests—which are often exceptionally threatened already through a host of other problems such as overlogging and slash-and-burn cultivation. Being stressed to begin with, the forest tracts in question will be all the more susceptible to injuries from acid rain. Conversely, acid rain on undisturbed forests will leave them vulnerable to greater damage from timber harvesters, slash-and-burn cultivators, and other exploiters of forest ecosystems.

A still bigger synergism arises with respect to public health. Already there is a powerfully compounding interaction between air pollution, bad water, malnutrition, and food shortages. This disease recipe will shortly be reinforced by hosts of new pathogens in a greenhouse-affected world—by which time people may well be finding their immune systems depressed as a result of increased UV-B radiation. The prospect is compounded by the fast-expanding factor of international travel by hundreds of millions of people each year—a pathogen's dream.

All too often, the biggest reinforcer of environmental damage is population growth. Consider the case of India and its use of coal for energy, contributing to global warming. With a per-capita GNP of only $350 per year, or one-sixtieth that of the United States, India's electricity capacity is about twice that of New York State. Although the country possesses meager coal reserves, it wants to exploit them fast, and to use much more oil too. During the past four decades India has expanded its coal burning six times and its oil consumption one hundred times. The government plans a number of energy-based initiatives for development—for instance, to supply electricity to half the country's homes. This measure, together with other development plans, will shortly induce a doubling of India's carbon emissions.[4]

At the heart of this prospect lies population, both its size and growth rate. India's population of 900 million people, growing at 2 percent per year, is projected to top 1 billion before the year 2000 and to approach 1.4 billion by 2025. Even with its low per-capita income and its less-than-advanced technological capacity, India's huge population makes for a disproportionately large potential contribution to global warming. Suppose, however, that India manages to reduce its fertility rate to replacement level (almost half the present number of children per family) within the next three to four decades; suppose also that at the same time it does no more than double its per-capita use of commercial energy (roughly matching that of China today). This increase, given the multiplier effect of the huge present population and its rate of growth, would result in India's emitting carbon dioxide at an annual per-capita rate of 1 ton by 2024, or roughly the world average in 1990. Because of the population factor, this would still be enough to more than cancel out the benefits of a putatively extreme step elsewhere, namely, the termination forthwith of all coal burning on the part of the United States without replacing it with any other carbon-containing fuel.[5]

When we consider all environmental disruptions together, and allow for the mutually amplifying effects at work between just some of them, we shall find there will be potential for exceptional numbers of synergistic interactions with exceptional impacts.

What does all this mean for security concerns? In view of the multiple and largely simultaneous impacts of these interactions during the coming decades, a single-problem-at-a-time account of effects will surely underestimate the eventual outcome overall. The phenomenon of synergisms means we should anticipate a greater environmental debacle, arriving more rapidly, than is usually predicted. In turn, it will mean a more profound disruption of the planetary ecosystem than we can presently foresee.

Fortunately there is an upbeat side to all this. A positive

synergism can arise through measures to tackle environmental problems. That is to say, a measure can, through intersector reactions, generate a compounded payoff. For instance, large-scale tree planting in the humid tropics, undertaken to generate a sink for atmospheric carbon dioxide and thus to counter the greenhouse effect, would supply many spinoff benefits. To the extent that the plantations would serve extra purposes that would not conflict with the carbon-absorption goal, notably commercial logging, they would relieve some exploitation pressures on the remaining natural forests—which could then continue to absorb carbon dioxide, and to foster other types of climate stability. In addition, reduced deforestation would serve to safeguard the uniquely abundant stocks of species in tropical forests, together with the gene reservoirs of these species—including, for example, a wild rice that through its disease resistance has helped save much of the Asian crop from a blight disaster. Both tree plantations and surviving natural forests could, in upland catchments, provide watershed services that are frequently degraded when tree cover is eliminated. There are a host of reasons why remaining tropical forests should be protected, indeed expanded. The greenhouse-countering rationale serves as a powerful reinforcing argument— which would carry little clout at all if the deforestation problem was not there in the first place.

So too with energy. To cite Amory Lovins, as Americans grow more efficient in the way they consume fossil fuels, they do several things at once. They foster greater economic growth for each unit of energy, saving money each year equivalent to half the federal deficit. They do wonders for trade and budget deficits. They reduce acid rain and city smog. They slow carbon-dioxide buildup in the atmosphere. Most important of all in the short run, they reduce U.S. dependence on Persian Gulf oil, with all the security implications that entails. Some synergism![6]

Probably the greatest example of these positive impact-reinforcing linkages lies with population planning. One of the best ways to control population growth is to reduce infant mortality, thus enhancing motivation for family planning. One of the best ways to reduce infant mortality is to supply better drinking water—which, via the watersheds connection, brings us back to tree planting. Plainly, the more that developing nations come to grips with population growth, the better they can deal with their other problems—as witness the success of nations as diverse as China, South Korea, Taiwan, Sri Lanka, and Cuba in rapidly bringing down their fertility rates, and also putting up a better record than most in across-the-board development.

All in all, the question of synergisms merits detailed and systematic attention if we are to comprehend all crucial factors at work in our environmental predicament and determine which of them are the most significant. If we fail to discern a synergism at work, our best efforts to tackle environmental problems may fall far short of what is needed. The same applies to constructive synergisms. So should we not consider a research agenda designed to confront the environmental challenges implied by synergisms? The amount of research under way is all too little, and it is almost entirely uncoordinated. Herein lies a major challenge (even a synergistic challenge?) for environmental scientists.

Environmental Discontinuities

There is a further side to the synergisms issue. When a synergized reinforcement takes place, there can arise what is known in the trade as an environmental discontinuity. It is better called a threshold of environmental injury, or a "breakpoint." It occurs when, for instance, growth in human numbers, in conjunction with growth in human consumption and growth in environmentally adverse technology, serves to build up a situation that can eventually generate an "overshoot" outcome.

In turn, this outcome can precipitate a downturn in the capacity of environments to sustain human communities at their previous levels. Technically speaking, the phenomenon occurs when ecosystems have absorbed stresses over long periods without much outward sign of damage, then eventually reach a disruption level at which the cumulative consequences of stress reveal themselves in critical proportions (an example is acid rain). We should anticipate that as human communities continue to expand in numbers, they will exert increasing pressures on already overburdened ecosystems and natural-resource stocks, whereupon environmental discontinuities will surely become more common.[7]

An instance has arisen in the Philippines where the agricultural frontier closed in the lowlands during the 1970s. As a result, multitudes of landless people started to migrate into the uplands, leading to a buildup of human numbers at a rate far greater than that of national population growth. The uplands contain the country's main remaining stocks of forests, and they feature much sloping land. The result has been a marked increase in deforestation and a rapid spread of soil erosion.[8] In other words, there occurred a breakpoint in patterns of human settlement and environmental degradation. As long as the lowlands were less than fully occupied, it made little difference to the uplands whether there was 50 percent or 10 percent space left. It was only when hardly any space at all was left that the situation altered radically. What had seemed acceptable became critical—and the profound shift occurred in a very short space of time.

This problem of land shortages is becoming widespread in many if not most developing countries, where land provides the livelihood for three out of five people and where much of the most fertile and accessible land has already been taken. During the 1960s, arable areas were expanding at roughly 0.5 percent per year. But during the 1980s the rate dropped to only half as much, and, primarily because of population growth, the amount of per-capita arable land declined by 1.9 percent per year.[9] Similarly, the annual expansion of irrigated lands, which supply one-third of our food from one-sixth of our croplands, fell by half during the same period.

So much for synergisms and the associated phenomenon of environmental discontinuities. It would be irresponsible to try to indicate what could be some of their impacts and their linkages to security, since we know so little about them—about their manifestations, let alone their workings and consequences. But the greater ignorance would be to ignore them.

The Unknown Unknowns

Still worse than to ignore the semi-unknowns would be to ignore the wholly unknowns. We face a host of "unknown unknowns." Recall that at the 1972 Stockholm Conference on the Human Environment, there was no mention of acid rain, global warming, and ozone-layer depletion, even though they now rank high on the environmental agenda. This raises a key question. What new unknowns are waiting in the wings to leap out at us? Could we not convert them into "known unknowns" in advance by at least identifying them, giving them a local habitation and a name? Or will we wait until we are surprised by them? What could be the "new global warmings" and other super-surprises?

This ranks as one of our biggest challenges of all. We are pretty good at analyzing problems when we recognize their existence. We are less skilled at reaching out to entirely new problems before they reach out to us. If we could do a better job on this front, we could often launch a preemptive strike and prevent many problems from becoming problems at all.

Where should we look for some of these unknown unknowns? What frontiers of environmental science should we be probing with a greater sense of exploratory foresight? Some people might protest that the endeavor is impossible by definition. We do not know what we do not know, so there we have it—or rather, we don't have it and we can't have it. But we could learn from our experience. A Cambridge University scientist, Dr. Joe Farman, announced the likely emergence of the Antarctic ozone hole in the late 1970s. His assertion was greeted with disbelief by fellow scientists. He must have made a mistake with his measurements; the atmospheric models of the time could not accept his calculations. It took several years before the ozone hole became all too apparent—years during which we could have done lots to limit the damage. When the "scientific wisdom" wised up to the real world, it got its arithmetic wrong again. It badly underestimated the speed at which the hole was widening and deepening.

If scientists can be so badly off target with respect to known unknowns, how will they ever anticipate what could turn out to be much greater problems, especially the synergized problems? As was pointed out in Chapter 2, we are conducting a planet-sized experiment with our environment, an experiment completely unplanned. Are we committing ourselves, or even condemning ourselves, to decades of unknowns? And along the way, to insecurity unlimited?

THE NEW

SECURITY

15

Trade-Offs
with Military Security

The arms race is actually killing people without the arms
being used.

—Willy Brandt, 1986

T he world is still stuck with
the idea that security is best, and often only, achieved through
military activities. This is revealed through the continuing
high level of military expenditures: almost $1 trillion in 1991,
or almost $3 billion a day and $2 million a minute. The annual
total is equivalent to the collective incomes of the poorest half
of humankind. True, there has been a slight decline in the total
since the end of the Cold War, roughly $100 billion per year.

But even if this recent decline can be maintained, we shall still have to wait six years before growth in per-capita GNP worldwide catches up with growth in military spending worldwide, seven years before the world's expenditures on arms and armies are no higher than they were in 1980, eighteen years before the world's arms trade drops back to the more moderate levels of the 1970s, and 125 years before annual education expenditures per school-age child match the current level of military expenditures per soldier.[1]

In the meantime we have almost as many wars and conflicts as ever, well over thirty of them in 1992. As this book has shown, a good number of them have their origins in environmental problems. So the question arises: could we buy more security—real, enduring, and all-round security—by diverting some part of military spending to environmental activities? Has the time arrived when we should be asking if we could be more secure if we were to forgo one additional super-sophisticated fighter plane and utilize the funds, $60 million, for tree planting, soil conservation, energy efficiency, species protection, and the like, plus population planning, poverty relief, and whatever else supports the environmental cause?

However valid these trade-offs may be in principle, the skeptic proclaims that is not the way the world works. Perhaps it should; but it doesn't. Political leaders and military strategists are simply not going to make choices along those lines. This realistic reaction notwithstanding, it is also realistic to recognize that political leaders and military strategists do make choices of a parallel sort. Concealed choices they may be, but effective choices they still are. They make them each time they determine how much of a national budget shall be allocated to military activities. Built into their choices is, willy-nilly, the unacknowledged fact that each dollar assigned to armaments means one dollar less available for clean water as a support to health, or for soil conservation as a support to agriculture, or dozens of other purposes. However little the budgetary pieslicers may be aware of it, they make choices in these terms effectively and increasingly. They do so with massive deprivation for non-military sectors, certain of which can contribute substantially to security in the proper broad sense. If choices are unavoidably being made and trade-offs are automatically being set up, why not do so explicitly rather than implicitly: by design rather than by default?

Specific Trade-Offs

To assess these trade-offs, let us look at some specific examples. What could be achieved if certain military expenditures were to be utilized for environmental purposes? Herewith is a short sampler of possibilities,[2] based on a figure for worldwide military outlays of $3 billion a day:

1. To boost developing-world agriculture to an extent that would allow all people to go to bed with a full stomach would cost $40 billion a year, or less than the annual cost of the Strategic Defense Initiative program in the United States. The same sum is spent each year on malnutrition of a different sort, through slimming aids to counter overeating.

2. To provide safe water plus sanitation for the one-third of the world population who don't have it (impure water is associated with 80 percent of developing-world diseases) would cost $36 billion a year. This is equivalent to little over twelve days of military spending.

3. To protect the ozone layer by supplying CFC substitutes and converting industrial technologies in developing nations would cost $1 billion a year, or eight hours of military spending.

4. The amount spent on development of new weapons, roughly $120 billion a year, is eight times greater than the amount spent on research into new sources of energy. Alternatively reckoned, it is more than all research into pollution controls, enhanced agriculture, and improved health as well as energy technologies. Note, moreover, that while new American cars are twice as fuel efficient as two decades ago, American weapons are one hundred times as efficient in accuracy and destructiveness.[3]

5. To supply a global five-year child immunization program that would prevent one million deaths a year (and thus help to foster family-planning motivation among developing-world parents) would cost around $1.5 billion, or the same as one Trident submarine or one day of the Gulf war. A child who is not immunized faces a thirty-three times greater chance of dying from disease plus malnutrition than of dying from war. To eradicate malaria, a killer disease that claims the lives of 1 million children every year, would cost $700 million, equivalent to six hours of military spending.

6. To save 2 million developing-world children who die each year from diarrhea, a result primarily of inadequate water supplies, would cost $50 million, or less than half an hour's spending on ways to kill people.

7. To provide family-planning facilities to all couples who want them would cost $3 billion a year (and thus reduce the ultimate global population by more than 2 billion people), equal to just a single day of military spending.

8. To reverse desertification would cost $12 billion a year, or four days of military spending. An anti-desertification campaign would also bring agricultural benefits of $30 billion a year.

9. To fund the Tropical Forestry Action Plan would cost $2 billion a year, or the equivalent of sixteen hours of military spending—also the equivalent of one nuclear-armed submarine.

10. To document the survival outlook of all the world's plants (tens of thousands remain undiscovered, let alone assessed for their potential against diseases such as cancer and

AIDS) would cost $1.3 billion, or the equivalent of two B-2 bombers.

Many more instances of environmental trade-offs can be cited.[4] In terms of human welfare generally, whether or not connected to environmental issues, note a trade-off that concerns one of the biggest blights of our world, malnutrition and associated disease. To reduce hunger among the 150 million poorest households would cost $6.4 billion a year. To cut malnutrition among women and children, $1.6 billion. To eliminate deaths from famine, $550 million a year. Total annual bill, $8.55 billion—or the equivalent of fewer than three days of military spending.[5]

In each case of environmental outlay, the amount in question is large in every respect except one, the comparison to present military spending. All items, plus many more to ensure an environmentally secure world, could be accomplished through the transfer of a tiny fraction of total military expenditure to environmental needs. Of course such an outright transfer is unlikely, to say the least. But more marginal trade-offs would be another matter. Political leaders might ask themselves whether each additional annual outlay of, say, $25 million for extra military hardware provides a greater incremental increase in security than could be accomplished if the funds were assigned to a host of alternative ways of promoting stability through environmental safeguards. Whereas $25 million represents only a mini-marginal expansion of most military budgets, it would make an absolute difference if applied to national budgets for, say, improved water supplies.

What would it in fact cost to fix up the biggest environmental problems, especially those of global scale, such as mass extinction of species, ozone-layer depletion, and global warming? Of course, these huge problems arise from other problems. Mass extinction of species occurs primarily in tropical forests, and can be best reduced by tackling tropical deforestation.

Global warming covers a host of activities that should be phased out for all kinds of other reasons, such as wasteful use of all energy in general and fossil fuels in particular. Supposing we come to grips with all environmental problems that have a global dimension of some sort, what would it all cost?

A lot and a little. According to Maurice Strong, a U.N. leader who organized the 1992 Conference on Environment and Development, the tab for cleaning up the more pressing international problems of the global environment would be around $625 billion a year. (This is still only two-thirds of military spending worldwide; couldn't the world be made secure through conventional means for a modest $1.7 billion a day?) Most of this overall sum would have to be supplied by individual governments to meet their own needs. But when it comes to those global problems and sub-problems that are largely located in the South, governments in question feel they could not readily pick up their "fair" share of the tab. It would be hard for them to sell the case that their citizens should cough up extra taxes to help protect everybody's tomorrow when one billion of their people have to remain focused on tonight's supper.

Strong proposes that the rich nations should contribute $125 billion a year to lend a hand to everybody's problems in the South. Large as this sum may sound, it is still equivalent to only seven weeks of the world's military spending. Or, to look at it from another angle, it would work out per Northern taxpayer to no more than $200 per year—or roughly half the average amount by which Northerners grow richer year by year. Or again, and as Strong puts it, after a highly successful career in the business arena as well as the environmental field, "Earth Incorporated is literally in liquidation. Much of the income we are producing isn't really income at all; it's running down our natural capital. If we were to provide $100 billion or two percent of the Earth's GDP for amortization and depreciation—that's much less than most corporations do—we would be able to put Earth on a sound, sustainable basis."[6]

Military Spending
in the South

To put the arithmetic into a little further perspective, let us take a look at who spends what on the traditional security front. Military expenditures are mostly confined to no more than one dozen countries in the North, whereas outlays on environmental support for, say, agriculture's environmental base would principally benefit more than one hundred countries in the South. Yet a number of developing-world governments engage in much military spending of their own (some of it by virtue of their role as client states of major Northern powers). Developing-world military expenditures have grown from $27 billion in 1970 to roughly $180 billion in 1991, accounting for almost one-fifth of the global total. Worse, their spending has risen at an average annual rate three times higher than developed nations during the last two decades, even though the average per-capita income in developing nations is only one-twentieth that of developed nations. Since 1960, developing nations have increased their military expenditures more than twice as fast as their per-capita income. Their 1990 outlays are the equivalent of $220 per developing-world family, or a whole one-quarter of per-capita GNP.

Military Spending
by the United States

As for the North, this is where the great majority of military spending has taken place. Today it stands at $780 billion, or more than four-fifths of the global total.

To be specific, consider the established stance of the United States, which today accounts for one-third of global military spending.[7] Since 1946 the nation has invested over $10 trillion in military activities, or way more than the value of all the country's tangible assets except the land itself. The great bulk of the investment has been directed to worst-case scenarios of the Soviet threat, even though the record now reveals that the scenario estimates were wildly over the top (by contrast with new environmental threats, about which certain military experts complain the threats are not demonstrated concisely enough). In 1990, the nation spent half as much again on its military as a decade back, and today the amount is only marginally less than at the height of the Vietnam War. Nor does the government plan any spectacular cutback. The Pentagon's 1991 five-year plan proposed that the defense budget in 1996 should be one-quarter more in constant dollars than in the mid-1970s when the Cold War was at its most frigid; and from 1996 till 2001 it should remain at the same level. Yet according to Robert McNamara,[8] a defense expert if ever there was one, it is eminently possible to cut the budget from 6 percent of GNP to only half as much in 1997. McNamara also reckons that if the whole world were to risk a mere 3 percent cut, within a decade that would release $1.5 trillion a year in constant dollars.

So much for the prospective peace dividend in the United States. Such funds as are made available will probably be spent for the most part on rebuilding American infrastructure (roads, bridges, and the like). A better target would be environmental infrastructure worldwide. Suppose the Strategic Defense Initiative, surely an item beyond its "sell by" date, were to give way to a Strategic Environmental Initiative as proposed by Vice-President Albert Gore and Senator Sam Nunn. Or suppose there were to be a new form of Marshall Plan, that is, along the lines of the visionary measure of the United States during 1947–52. That plan transferred $110 billion in 1991 dollars to Europe, the biggest aid program the world has ever seen and surely the soundest investment the United States has ever made. It amounted to 2.8 percent of GNP. The United States' foreign aid today is less than 0.2 percent of GNP even though the economy is three times as big. Yet one billion people in the developing world are in a much worse plight than the people of war-ravaged Europe, and the United States' stake is no less critical. The problem is not shortage of money, it is shortage of vision.

16

The Policy Fallout

Those who make $200 a year should not pay so that those who make $10,000 a year can breathe clean air. We are all in the same planetary boat. A few of us travel first class, while most are in steerage. But if the boat sinks, we all drown together.

—Edward Kufuor, chairman of the Group of 77
(the developing nations' "club"), 1991

To reiterate this book's theme, security interests are under siege from environmental threats. These non-military threats can be met only by non-military responses. An artillery salvo does not serve against soil erosion, rockets will be unguided when the opponent is acid rain. So how shall we start to devise responses that are on target? In particular, what should be the policy strategy to reflect the new challenges? What policy levers can be pulled by political leaders?

Let us consider two broad dimensions of the situation overall.

Environment and International Relations

First off, we need to take a long look at the fast-changing character of the global community now that it has to share the stage with environmental threats. The new scene represents an upheaval for international relations. For instance, no nation can now go it alone in a world where pollutants, whether acid rain or greenhouse gases, recognize no frontiers. This amounts to as great a change for the nation-state system as any since the emergence of the nation-state four centuries ago.

So those with their hands on policy levers need to take a firm grip on an entire array of fresh factors. Without a clear-eyed view of these factors, political leaders will simply not know how to rise to the challenge of countering soil erosion, decline of forests, and spread of deserts, let alone climate change.

But straightaway they run up against a major road block. Environmental issues are exceptionally interwoven in both their causes and their consequences. So political leaders need to grasp not only the entire range of accumulating problems but their multiple interactions as well. Many of these interactions appear set to grow still more numerous and significant and to exert an ever greater impact.

Regrettably, environmental problems do not often lend themselves to ready and precise analysis. The interlocking webs of cause and effect do not allow us to treat the issues in the way we are accustomed to when dealing with major issues, whether national or international. We cannot generally say that environmental problem X will necessarily lead to economic consequence Y that demands policy response Z. It cannot be predicted precisely how these factors will interact among themselves. Still less can we figure out the specific ways in which they will interact with other factors, such as the growing tools of violence available through the arms trade, the spread of terrorism, and the proliferation of weapons—biological, chemical, and even nuclear—to nations and dissident organizations (the "backpack device" may come on stream by the year 2000). But it is realistic to suppose that environment-based tensions and conflicts, whether domestic or international, will escalate and multiply. The most remote eruptions could readily reach back to the United States by both human and natural means. Confucius had the idea 2500 years ago: "If a man take no thought about what is distant, he will find sorrow near at hand."

Our difficulty in recognizing connections between the environment and stability may say less about the nature of the connections and more about our limited capacity to think clearly about matters that are new to our experience. Even the connections that are more readily apparent do not always lend themselves to quantification, in contrast with "conventional" connections, such as trade and investment flows, whose effects are easily measured and thereby carry greater weight with policy experts. What can be counted should not be emphasized to the detriment of what also counts.

The "difficult to discern" linkages between environment and security may well explain why the injunctions of the Brandt Report[1] have been largely ignored since its release as far back as 1980. To quote:

> Few threats to peace and survival of the human community are greater than those posed by the prospects of cumulative and irreversible degradation of the biosphere on which human life depends. In a global context, true security cannot be achieved by mounting buildup of weapons (defense in a narrow sense), but only by providing basic conditions for solving non-military problems

which threaten them. Our survival depends not only on military balance, but on global cooperation to ensure a sustainable environment.

Despite this resounding assertion of the environmental underpinnings to security, plus many more statements by such leaders as former president Mikhail Gorbachev, Chancellor Helmut Kohl, President François Mitterand, and former prime minister Margaret Thatcher, there has been next to no effort to translate the principles into practice. This may reflect the difficulties that policymakers encounter when trying to determine the specific trade-offs in question. But pragmatic instances abound.

Consider the American experience again. In the case of Egypt, the United States in 1986 supplied $1.4 billion in what is technically known as security assistance and $1.3 billion in other forms of security support. Only a small part of this aid was directed at environmental measures to safeguard Egypt's overstretched capacity to feed its fast-growing population. A key question arises. Which of the two, military or environmental outlays, would purchase more real and enduring support for U.S. interests in Egypt and, because of Egypt's international role, in the Middle East and the Arab world generally?

What if, moreover, a good number of other countries are steadily overwhelmed by environmental problems, leading in turn to economic and political problems? Envision a future where these problems grow more numerous and widespread. The overall outcome could be a continuing downward spiral of poverty in many developing nations, associated with deficits of food, energy, and water, plus urban squalor and widespread unemployment. In turn, these factors could precipitate economic stagnation, social frustration, political upheaval, and even government paralysis or collapse.

What happens when the negative cycle takes hold in nations that are important to the United States by reason of aid, trade, investment, political affinity, strategic location, and other vital interests? The United States could find itself increasingly beset by requests for further aid and humanitarian help. It could even face "economic aggression" in the form of trade cartels, market closures, nationalization of assets, and debt cancellations. Were the governments of beleaguered nations to seek to relieve their domestic tensions at any cost, they could become increasingly inclined to overexploit their environmental resources in order to meet immediate threats. Further, they could become characterized by autocratic and repressive regimes and build up their military so as to contain internal conflicts or resort to excessive international assertiveness to gain leverage abroad. If many other developing nations followed suit, widespread political instability could severely undermine the vital interests of the United States.

We cannot predict exactly how these factors will interact, or how far—in light of the global traffic in arms, the prevalence of terrorism (remember the World Trade Center), and the proliferation of nuclear devices—they will lead to violence. But it is realistic to suppose that tensions and conflicts, whether domestic or international, could easily increase and intensify. They could even destabilize the emergent new world order.

However multifaceted, subtle, and difficult to discern are these linkages between developing nations and the United States, they are real and significant—and they are growing more numerous and influential. On top of all this are the more direct and still greater problems of ozone-layer depletion and global warming, even more threatening to the United States. Fortunately all these problems are amenable to remedial measures provided they are recognized in due time and with focus on their true extent. The question at issue is whether the United States will recognize its interests in their proper scope, and respond accordingly. The United States will have to respond sooner or later: either sooner, through cooperative measures of sufficient scope, or later, through measures to grapple with the problems of an indivisibly impoverished world.

The New Kid on the
Global Block: Interdependence

To reiterate a basic point, no nation can shield itself from diverse forms of environmental degradation in other nations. Not even the United States, which is economically the largest, technologically the most advanced, and militarily the most powerful nation on Earth, can isolate itself from the many environmental problems, together with the security threats they pose, in other parts of the world. All nations are in the same boat, and it is becoming an environmental *Titanic.* The name of the new game, interdependence, may be cumbersome to the tongue and eye, but it is one we shall all have to learn to speak easily and hear readily, just as we all practice it throughout our daily lives—lives in which interdependence has become a fact of life.

Note, for instance, that transfrontier financial dealings each day amount to $81 trillion, or three times more than the annual Pentagon budget, equal to almost one-twentieth of the global economy. This means that one nation can effectively export prosperity and stability to other nations, just as it can export inflation and unemployment too. In turn again, it means that government of our affairs is increasingly undertaken by bodies other than governments—bodies that reach across political boundaries and operate primarily through international networks. Anyone who doubts this should recall the Wall Street crash of 19 October 1987 ("Black Monday") when the add-on effects reverberated throughout the world's financial system. There is no longer a New York Stock Exchange and ditto in London, Frankfurt, and Tokyo. There is a single global system. What price (so to speak) the archaic ideas of national independence?

Interdependence implies a growing premium on collective action to tackle collective problems. As is demonstrated by the example of greenhouse-gas buildup in the atmosphere, there are environmental problems to which all nations contribute and by which all will be affected. Similar considerations apply, though on a more limited scale, to such problems as ozone-layer depletion and mass extinction of species. So while there is now less leeway for a superpower such as the United States to "go it alone," there is greater scope for the nation to cooperate with other nations. In turn this postulates a policy mind-set that reflects the new constraints—the creative constraints—of interdependence.

Accepting the challenge of interdependence means recognizing that many environmental problems—and the opportunities they imply, too—constitute a distinctive category of international issues, unlike any of the past. The new agenda lies beyond the scope of established diplomacy and international relations, and requires a new emphasis on enlightened self-interest of both individual nations and the whole community of nations, expressed through an unprecedented degree of international cooperation.[2]

Interdependence also means an adjustment of many other relationships, especially between the developed world and the developing world. Whereas the past forty years have been dominated by tensions between West and East, these are now being replaced by confrontations between North and South—confrontations that show every promise of extending another forty years unless the tensions can be swiftly defused. Fortunately they can be defused all too readily because there will be benefit to both sides (as invariably happens with interdependency relations anyway). Competitive security must now give way to cooperative security.

Until yesterday, one side in a dispute could feel strong only in relation to the other's apparent weakness. The same applied the other way around. In response, both sides would feel driven to escalate their military strength, whereupon both would drain their economic strength in an unending struggle for supremacy of a conventional sort. As a further result, and to cite the late Professor Kenneth Boulding of the University of Colorado, the greatest threat to national security has become national defense. How different is the situation today, when both sides will win together or both will lose together. This is a transformation from the traditional eyeball-to-eyeball stance that has been the norm ever since people took up arms. Now the arms required are those of an embrace. No wonder that political leaders find it hard to climb into the new mind-set.

The North–South Confrontation

As noted, the principal "big picture" scene of international tension today is between nations of North and South. As long as this persists, there is all the less prospect of a joint hands-on effort to tackle global environmental problems such as mass extinction of species and global warming. What is the source of this confrontation? Let's check a few items, first as seen by the South:

1. International Debt

Developing nations owe $1.2 trillion of outstanding loans to developed nations. To pay the annual interest and a little of the capital, they fork out more than $150 billion a year to their Northern creditors.[3] This debt burden is much greater today than it was ten years ago because of vagaries in the currency through which debt is handled, the U.S. dollar. When President Reagan embarked on his military buildup and pushed the U.S. federal deficit to new heights, the strain on the American economy caused interest rates to soar. Every time Wall Street registered a one-point rise during the 1980s, Brazil's debt went up by $4 billion. In vain do Brazilians and other debtors protest that they should not suffer economically because of military activities in a Northern country. Meanwhile, the more they pay, the more they owe.

The debt burden weighs heavily on the South's development generally[4] and on its environments in particular. It has induced governments to overexploit resource stocks such as tropical forests in order to generate immediate, albeit unsustainable, revenues.[5] It has obliged many poor nations to cut back on their government spending on health and family-planning programs, and has thus contributed to the slowing in fertility-rate declines in the Philippines, India, Tunisia, Morocco, Colombia, and Costa Rica.[6] It can even be held responsible for causing the deaths of half a million children each year by slowing development,[7] with all that implies for population-planning prospects.

During the second half of the 1980s, there was a flow of funds from North to South, comprising foreign aid, World Bank loans, and the like, totaling roughly $90 billion per year. But when the calculus includes debt repayments, it turns out that there was a net flow of around $40 billion from the South to the North. It was akin to a blood transfusion from the sick to the healthy. It also induced some economic sickness in the North. The debt burden has limited economic growth in the South, and thus reduced the South's capacity to purchase the North's exports. The cost to the United States has been calculated to have entailed $60 billion of sales forgone and the loss of 1.8 million jobs.[8]

2. Aid

It may seem strange that the South should feel in a position to criticize the North over aid. Even if certain nations of the North have reneged on commitments time after time, should not the South be duly grateful for what it does receive? That depends on how we see aid. Aid is one of the best investments available for nations of the North. Every one dollar handed over in aid brings back an average of two dollars in increased trade, apart from other returns such as satisfaction in assisting with humanitarian endeavors. It should not be donated out of a spirit of charity; rather it should be dispensed out of a sense of joint responsibility for an interdependent world. Otherwise aid demeans those who give it and degrades those who receive it.

A decade and a half ago the community of nations, North and South, agreed that the amount of aid should be 0.7 percent of Northern nations' GNP. Since that time, most Northern nations have slipped back. In 1990, total aid amounted to $54 billion, or 0.35 percent of the North's GNP. The United States, then the biggest source of aid, although since overtaken by Japan, provided $11 billion, 0.21 percent of the U.S. GNP. My own country, Britain, was little better: 0.27 percent of its GNP. Several small nations with big ideas led the field: the Netherlands, Denmark, Sweden, and Norway were all around 1.0 percent.[9]

Think of what Americans with their wealth could do. An annual 2 percent increase in their affluence works out to an extra $450 per citizen, or twice as much as the entire annual income of each of the 400 million poorest people. If U.S. aid were to be tripled so that it approached the 0.7 percent level, it would cost Americans the price of a beer a week. Yet $35 billion would achieve much for developing nations. It would be equivalent, for instance, to one-quarter of the collective GNP of sub-Saharan Africa.

3. Trade

Third, international trade has an adverse impact on environmental concerns. Agricultural trade protectionism on the part of developed nations in North America, Western Europe, and the Pacific entails the outlay of $200 billion a year to protect their farmers. The policy works against agricultural exports from developing nations, depriving them of trade revenues worth $30 billion a year.[10] In turn, these direct losses reduce developing-country farmers' profits, leaving them less money to invest in upgraded agriculture and thus perpetuating poverty. In turn again, they ultimately induce poor farmers to overload their croplands and to encroach onto marginal lands,

lands that are too wet, too dry, or too steep to sustain agriculture.

More important still are trade barriers and other forms of protectionism in the North—a restraint that seems to be on the rise in Washington, London, Bonn, and Tokyo. President Clinton, Prime Minister Major, Chancellor Kohl, and others proclaim the virtues of free-market competition, but they have hitherto confined it for the most part to their own sector of the world. When it comes to developing nations, their response is, "Sorry, come back another day." Were South–North trade to be liberalized rather than limited, the South would benefit by $100 billion a year.[11] Northern consumers would receive a windfall, too. The average American would pay $350 less per year for imported goods of similar quality and lower price than American products.

On top of these three leading items, there are other ways in which, to some extent at least, the poor stay poor because the rich are becoming richer. For instance, commodity prices have steadily declined over the decades. A Kenyan farmer wanting to buy a small tractor from Detroit must now sell four times as much coffee as thirty years ago.

When we include the entire range of factors that deny developing nations access to international markets and the like, we find the total swells to $500 billion a year,[12] or almost ten times more than foreign aid. With funds such as these—they are equivalent to one-fifth of developing nations' economies—the South could do much more to safeguard its environments, its economies, and ultimately its all-round security.

Next, the standoff as seen by the North:

1. Population Growth

This is often at the top of the list of Northerners' complaints. As we have seen in Chapter 10, it is a serious concern for all. The issue presents lots of good news, as witness the remarkable family-planning breakthroughs in Thailand, Indonesia, Sri Lanka, and Cuba, among a dozen other countries. There is also a plethora of dire news in those many countries which have still to be convinced there is much of a problem at all. The situation would be helped enormously if the United States were to resume its out-front role in family planning. Since the early 1980s there have been no American funds for the two main family-planning organizations on the international scene. One result is that there has been an upsurge in abortions in developing countries till they now total somewhere around 50 million a year, or two-fifths of all Southern births. This is an ironic outcome insofar as the American stance owes much to its anti-abortion lobby. As the ecologist says, you can never do only one thing.

In addition, let us bear in mind that the population explosion—or to give it a better name, implosion—is not just a case of increase of human numbers, but of consumer appetites too. The United States' annual increase is only one-quarter less than Bangladesh's, but because of Americans' use of fossil fuels it will cause thirteen times as much damage to the global climate as will Bangladesh's.

2. Off-Target Policies

Many nations of the South have put together a scrambled egg of policies. Through excessive subsidies, tax holidays, and other inducements, commercial loggers in Philippine rainforests have been making a superpacket while the economy has been losing $250 million per year, or five-sixths of what it could have earned. The result is massive overlogging and an end to the country's forests before the end of this decade—and an end too to hardwood-export revenues that once brought in a full one-third of the country's foreign exchange.[13] Much the same applies to cattle ranching in Brazilian Amazonia. (But while being critical of the South's policy mistakes as concerns its forests, let's recall that the rainforest being destroyed fastest, and largely through government subsidies, is the United States' Tongass Forest.)

The South could do much to get its policy house in order. It tends to favor manufacturing over agriculture, cities over countryside—the reverse of what generally works for long-term development. Nor will the South come to grips with imperatives such as a fairer division of farmlands, economically efficient and socially equitable at the same time; with frying off the fat from the bureaucracy; with exchange-rate overhaul; and a dozen other policy priorities.

3. Corruption

President Mobutu of Zaire is reputed to be worth more than Zaire. Many other developing-nation leaders are skilled at salting away funds, especially international funds such as foreign aid. (But before pointing a critical finger, let's bear in mind the Boesky scandal in America and the Maxwell roguery in Britain.) On top of this grand-scale corruption, there are other and more legitimate forms of "capital flight," including the simple transfer of funds from the South to safe havens in Switzerland and other Northern countries. These transfers are roughly reckoned to total $50 billion a year,[14] or much the same as all foreign aid in the other direction. The situation could be helped by the North through tighter regulation of their banking systems. Regrettably this is resisted by those small yet powerful sectors in Northern economies, the bankers' fraternity, who would lose a little business.

Were the South to tackle these problems, it would again put much money into its citizens' pockets. That is obvious. What is less plain is that the South would also, via numerous linkages whether direct or indirect, help to safeguard both its environments and its economies. There could be no better way to ensure its security.

On top of these six items, three on each "side," there are two further protests that relate particularly to the environment. The North considers that the global environmental crisis arises mainly from pressures on natural resources on the part of developing nations, notably topsoil, drylands, forests, montane areas, coastal ecosystems, and species. Conversely, the South sees the crisis as stemming primarily from excessive consumption of natural resources on the part of developed nations, notably fossil fuels and other key minerals, plus the capacity of the global atmosphere to absorb greenhouse gases and ozone-depleting chemicals. Each side blames the other with resounding rhetoric, ostensibly oblivious to the fact that when natural resources of global importance are misused and overused, it is the entire global community rather than separate sectors of it that will be indivisibly impoverished. For either side to point a critical finger is like proclaiming that the other side's end of the boat is sinking.

The Mediterranean Clean-up Plan

Fortunately, these North–South tensions can be balanced by a good-news item whereby nations of both sides have collaborated to splendid effect.[15] The Mediterranean Sea is used as a cesspool by the 100 million people who live along its shorelines and by the still larger throngs of people who take their vacations there each year, making up one-third of all tourists worldwide. As an enclosed sea with an outlet confined to the narrow Straits of Gibraltar, the Mediterranean renews its waters once every eighty years. Tides are small and currents weak, leaving pollution pretty much where people dump it.

Few areas of the world are subject to as much human disruption as the Mediterranean. It is one of the world's main waterways for shipping, with one-third of total petroleum trade, and annual oil spills plus tanker flushouts are seventeen times as great as that from the *Exxon Valdez.* But at least 85 percent of pollutants stem from land sources in the form of industrial waste, municipal sewage, and agricultural residues. Seventy rivers, large and small, daily deposit thousands of tons of industrial effluents. At least 120 coastal cities and towns pump 90 percent of their sewage untreated, or at best poorly treated, into offshore waters. Covering only 1 percent of the Earth's ocean surface, the Mediterranean contains one-half of all floating oil, tar, and general garbage that mess up the Earth's seas.

The region is subject to several endemic diseases, including viral hepatitis, dysentery, typhoid, poliomyelitis, and cholera. Periodic outbreaks of typhoid put dozens of people in the hospital. Along France's fashionable Côte d'Azur, sunbathers are warned off stained sands by pollution flags, and swimmers are kept out of the most fouled waters by police patrols. In 1979, nineteen people in Naples died of cholera after eating contaminated mussels—and in Rome, as in Naples, you hear that if you order oysters in a restaurant you are playing "Italian roulette." At the 1992 Olympics held in Barcelona, the offshore waters were so polluted that sea-sports competitors had to take special health precautions.

The Mediterranean harbors hundreds of fish species, many of them exotic enough to rank as luxury food items. While the annual catch amounts to only 2 percent of all fish taken around the world each year, its economic value of over $1 billion amounts to 6 percent. But now that mounting pollution is aggravating a history of overharvesting of fisheries, several species have declined in just a few years from exceptionally abundant to almost extinct. Consumers in Spain, Yugoslavia, Greece, Turkey, and Israel complain that the price of fish on their dinner plates is several times higher than that for Atlantic fish.

In addition to pollution, the Mediterranean suffers from poorly planned tourism. Hotels, marinas, and other facilities are desecrating one natural area after another and disrupting wildlife communities around the basin.

In 1975 the United Nations Environment Programme (UNEP) started on what seemed an absurdly ambitious project. It wanted to persuade the eighteen coastal nations to formulate a joint strategy to tackle the problem. In the event and through some fancy diplomatic footwork, UNEP succeeded in getting seventeen of the nations (the absentee was Albania) to sit down and formulate a plan of action. UNEP's feat appears all the more remarkable in light of some of the parties making common cause with each other: Israel and Syria, who are in a state of perpetual hostility; Egypt and Libya, long-standing foes; Turkey and Greece, with hundreds of years of enmity; France and its former colony Algeria, still mistrustful of each other; and Spain and Morocco, wary of ancient antagonisms.

Since 1973, there had been a protracted process of scientific research, economic evaluation, and legal planning. The upshot was an environmental breakthrough in late 1980, when a conference gathered in Athens. The participants agreed on a draft treaty, the final form of which came into operation in 1982. The principal output of the conference was a set of initiatives to tackle pollution. First of all there was a "black list" of contaminants that would steadily be eliminated from the scene by all source countries, especially by the three worst polluters, France, Italy, and Spain. Included on the black list were mercury, a metal highly toxic even in trace amounts, 100 tons of which are dumped into the Mediterranean each year; radioactive materials, of which 2500 curies of radionuclides were entering the sea each year; and a number of carcinogenic and mutagenic substances. A second "gray list" included substances such as lead, zinc, copper, titanium, crude oils and hydrocarbons, pathogenic microorganisms, nonbiodegradable detergents, and other substances including pesticides that were having an adverse effect on fish and shellfish (tuna, swordfish, and marine mammals contained five to ten times more heavy metals than their counterparts in open oceans).

Since these gray-list substances were less poisonous than the black-list items and were more easily rendered harmless through natural processes, some of them continued to be discharged into the Mediterranean—but only under new scientific control and licensing procedures. A broad-scale monitoring exercise was launched drawing on the coordinated efforts of dozens of laboratories in almost all countries.

Today the Mediterranean is somewhat safer for local residents and tourists. According to a 1982 statement by Dr. Stjepan Keckes, a Yugoslav marine scientist who headed the UNEP team, "A lot of the Mediterranean might look clean, but there is potent pollution from heavy metals and bacteria. While it is an illusion to imagine that the Mediterranean will ever be pristine, we have reversed the tide of pollution and eventually we should be able to guarantee safe, clean waters. Naturally this won't be done overnight. The skies of London were not made fog-free, or the River Thames safe for salmon, in a month or a year. But while the Mediterranean is sick, it is not yet dead. I believe we can make it a great deal better within a decade."[16] Here we are, a decade later, and the Mediterranean is a lot better indeed. Not healed by a long way, but no longer dying.

The treaty also set up an expanded network of parks and reserves for wildlife, both in the sea and on land. The present handful of protected areas will eventually be increased to more than one hundred.

The original agreement went way beyond the most optimistic expectations of observers, whether scientists, industrialists, or politicians. In terms of environmental politics among the community of nations, it ranked alongside a disarmament agreement. There could hardly have been a region of Earth with greater political disparities among nations in question, yet they were persuaded to rise above their individual interests in favor of the collective welfare.

The clean-up program has not come cheap. Within its first fifteen years it was expected to cost at least $15 billion to control pollution alone. But the nations concerned could not afford to turn away from the price tag of their past delinquency. The tourist industry alone was worth $10 billion a year, and the flood of sun-seeking visitors was projected to double by the year 2000. Eighty-five percent of the funds were to be supplied by the three countries most to blame, France, Italy, and Spain, these also being the nations with the biggest tourism industries.

The Mediterranean blueprint is serving as a model for parallel programs in other regional seas.[17] Twenty years ago the Baltic was so fouled that its fisheries virtually came to an end, but with sufficient political commitment and ecological know-how there is hope today that they can be restored to life. The Caribbean features more rapidly growing coastal industries, especially in the way of super-polluting petrochemical complexes, than any other marine zone and other nations. There may even be the prospect of a political-environmental breakthrough in that region of extreme discord, the Persian Gulf. As an indication of the problems involved, it took four years to achieve even preliminary agreement between the two main "sides," Iran and the rest. Part of the squabbling arose over the mere name. The word "Persian" did not suit the Arabs; a switch to "Arabian" was strictly unacceptable to the Iranians; and "the Gulf" was intolerable all around. So for participant parties, the zone was designed simply the Region. Yet so keen were coastal states to achieve accord that delegates from Iran and Iraq got together around a conference table and for several years they greeted each other with fraternal assurances, even while their two nations were at war. In the wake of the 1990 war, conservation plans have been shelved. But all nations involved know that it is in their joint interest to work together. One day they will start talking again.

The UNEP Regional Sea's Programme must count among the most remarkable of the agency's diverse activities. It accounts for a trifling part, less than one-tenth of UNEP's total budget, yet it has been advancing with giant strides. Governments have agreed to Action Plans for clean-up programs in seven regional seas around the world, and only three others remain to be tackled.

American Leadership

In conclusion, let us consider the expanded leadership role that is available to the United States. Were the United States to take steps to safeguard its interests in developing nations, it would need to engage in a growing array of collaborative endeavors. As more and more nations come to recognize their interest in the common global environment, the attack on collective problems through collective effort should gather momentum. This approach of "We're all in it together" places a premium on leadership—an ingredient that is all the more significant in light of the preeminent scientific skills, technological know-how, political acumen, and financial resources of the United States. The last item, reflecting America's economic prowess, is particularly pertinent. In any case, it need not cost the Earth to save the Earth. Indeed, because of growing interdependency relationships between the United States and developing nations, support for developing nations will prove to be a singularly profitable investment all around.

When faced with crises, policymakers can do more than react. They can turn crises to advantage. As old presuppositions and the policy regimes they supported are overtaken by events, the process opens up new opportunities for exceptionally creative initiative. In the past, this "all options are open" prospect has usually been available only in the wake of a major war when the old order has been blitzed and there is space for "start out afresh" thinking. At the 1945 conference in San Francisco that led to the foundation of the United Nations, the world was recovering from a catastrophe that had threatened civilization itself.

We need a similar spirit today. Nothing less will serve our present needs, even though we must do our postwar planning without fighting the war first. Or, to put it another way, we have to take the postwar steps while in the midst of our war against our environments. The time is surely ripe for us to be adventurous and head for new horizons. What may now appear politically impossible could soon become politically imperative. American leadership could generate much policy leverage. It has certainly done so in the past. The 1972 Conference on the Human Environment was instigated by American initiative. It was followed during the 1970s by a series of United Nations conferences that tackled water, deserts, tropical forests, and energy, together with associated issues such as food and population. In all these, the United States was generally out front, showing the way ahead through the creative flair and generosity of spirit that has often characterized Uncle Sam.

Will America's politicians now stand tall enough again that they can see beyond traditional horizons? The biggest problems are not scientific or technological or even economic. They are political, so they are questions of what we choose to do. By definition, success is ours for the taking. As was pointed out at the beginning of this book, our war against our environments is not a war of people against people or nation against nation. It is essentially a war against the old ideas of what makes our world turn.

Hard as it is for political leaders—or every last one of us, for that matter—to wrap our minds around the challenges ahead, let us take heart from some political advances in recent years that rank with the most basic in living memory. The Berlin Wall has come down and Germany is united. The Soviet Union has faded away. South Africa is coming to its senses. The Middle East antagonists are talking to one another. All these seismic shifts have required the elimination of walls in people's minds. Now we must move up a gear as we try to tear down the iron curtains in our minds.

In any case, change will come whether we positively pursue it or not, whether it arrives by design or by default. The option of business as usual was foreclosed many years back. Fortunately there is still time—though only just enough time—to choose the courses preferred. Through urgent and incisive action, the United States can generate a sizable payoff for itself as it moves to safeguard its stake in the global environment, whether in developing nations or in the rest of the world, including its own backyard.

The same applies to Canada, Britain, Japan, and all other developed nations. However powerful and preeminent they might think themselves in conventional senses, they can no longer stand apart from any other nation on Earth.

The Ethical Dimension

It is fitting to conclude with what could eventually turn out to be the biggest policy determinant of all. In Chapter 1 I noted that a prime motivation for the developed nations to support developing nations lies with humanitarian concerns. This potent issue reaches beyond real-politik factors. It postulates that we exist not only as members of individual nations and communities, but as participants in the common human enterprise that encompasses people all around the world. "Seek not to know for whom the bell tolls, it tolls for thee" (John Donne). This is all the more pertinent for the rich nations in that, being some of the most affluent societies that have ever existed on Earth, they bear a responsibility for the disadvantaged majority of humankind.

There is a further ethical aspect to policy. It reflects the long-term repercussions of our environmental assaults on our planet. We are currently polluting our ecosystems on a scale that, if the pollution were to be suddenly terminated in, say, the year 2010, the length of time needed for natural processes and human efforts to make good the damage would certainly be several decades. If we were to stop the spread of deserts in the same year, it would take at least a century to push them back again—and the same to allow the ozone layer to recover. Soil erosion would require many centuries before soil stocks could be replenished, and much the same would apply to tropical deforestation and global warming. All these restorative efforts would be needed to compensate for environmental injuries imposed within half a century. On the ethical front, we are all undeveloped nations.

In effect, leaders and citizens of the present generation are making decisions that will profoundly affect many generations to come. No human community in the past has wielded such a capacity to impoverish communities that will follow. Nor has any community of the past possessed such a capacity to safeguard the planet at a time of unprecedented threat. We still have time, though only just enough time, to turn a profound problem into a glorious opportunity.

17

Finale:
A Personal Reflection

War is often thought of in terms of military conflict, or even annihilation. But there is a growing awareness that an equal danger might be chaos—as a result of mass hunger, economic disaster, environmental catastrophes, and terrorism. So we should not think only of reducing the traditional threats of peace, but also of the need for change from chaos to order.

—Willy Brandt, 1986

And so to an end. Clearly, environmental security has become part of the established order of things, whether we all recognize it or not. It offers the prospect of an entirely new way of running our world—and equally, of how we view our Earth.

Recall the opening sentence of the report of the World Commission on Environment and Development, "Our Earth is one, our world is not."[1] Proclaimed six years ago, it is little heeded.

The Rio Earth Summit in June 1992 gathered together 170 nations, more than had ever assembled around a conference table. Almost 120 heads of state turned up, another first by a long way and 100 more than at the 1972 United Nations Conference on the Human Environment. They argued and negotiated, and produced piles of recommendations that reflected primarily the old order of individual nations with their individual interests. Hardly a single leader rose above parochial concerns to speak up for the Earth or the world. One year later the global community is less secure than ever. Its environments continue to be despoiled if not devastated on every side. The overdeveloped nations keep on consuming resources like there is not a this evening, let alone a tomorrow. Much of the underdeveloped world looks less likely than ever to develop, and poverty grows wider and deeper. At the same time, there are almost 100 million more people than at the time of the Rio conference.

No wonder that conflict and violence thrive, whether in Eastern Europe and the former Soviet Union, or in the Middle East, or in Central America and Southeast Asia, or in the forestlands of Peru and the Philippines, or in the shantytowns of dozens of Third World cities. The arms trade prospers, national leaders puff out their chests, and more nations are acquiring weapons of mass destruction.

All these thoughts wandered through my mind during an occasion at a remote-sensing conference in São Paulo, Brazil, a year ago. I noticed a wall map of the Earth, made up of satellite images taken by Landsat. It all looked compact and solid: dear old Earth. It looked just as one would expect, with those comfortable expanses of land among the dominant oceans. Fully familiar—except for something that snagged my attention, something far from familiar. I looked again. I saw. There were no political boundaries. The Earth was one all right, and so was the world. Then I reflected that those political boundaries were a major reason why the deserts were bigger than they would have been on a map half a century ago, and why the tropical forests were smaller. When nations preoccupy themselves with supposed threats next door, they overlook the greater threats at home, and they spend grotesque sums on the first threat to the detriment of the second.

Next to this satellite-image map was another space-agency picture, this one taken a quarter century ago. It was a shot from the Moon, showing Earth hanging in space. Another familiar sight. As I gazed at it, the thought occurred that no matter how many times I had looked at it before, I had not really seen what I was looking at. What the picture told me was that we are conducting our planetwide experiment far more energetically and ingeniously than when the shot was taken. Today we are acting for all the world (so to speak) as if we have a spare planet parked out there in space to which we can move if we ever find we have overcooked things on this one.

And the winds carry no passports. No more than rivers or birds, which likewise tell us that human-made divisions of our world are too artificial for the Earth under intolerable strain. Has the time arrived when we must finally heed the message that we can have peace on Earth only through peace with the Earth—and peace with the Earth only through peace with each other?

Some skeptics might still say the connections between environment and security are not apparent enough. To which I respond, "So what?" There are lots of other good reasons why we should be looking after our environments anyway—whereupon, as we have seen in Chapter 11 on global warming, we will find ourselves in a "no regrets" situation where we cannot lose. As that prolific writer Anonymous has put it, "If we live as if it matters, and it doesn't matter, then it doesn't matter. If we live as if it doesn't matter, and it matters, then it matters."

So in this final chapter, we come full circle. Environmental security: we can have it because we choose it or because our Earth eventually demands it of us. If we choose it, we will not find it all that expensive. It need not cost the Earth to save the Earth. The real question is not "Can we afford to do it one day?" It is "How can we not afford to do it right away?"

If we choose to choose, we shall find it will open up scope for security in the proper broad sense of the term—and security more abundant than we could hardly hope to envision right now. When the monies of the military are finally turned to the purposes of peace, there could be peace aplenty.

Spare a Thought

Let us think, then, of the future that is struggling to be born. At the same time, let us think about what we also do, without thinking about it, to preempt that future. Spare a thought about, for instance, the number of times a day we inadvertently do something that pushes the cause backwards. Like drive the car, 2000 pounds of steel, to transport just one person, and thus send a silent message of support to Detroit and its gas-guzzler fixation. Or generally overindulge our sky-high affluence. Or forget to write that environmentalist letter to Capitol Hill.

All those with life-styles that are simply not sustainable (like burning up fossil fuels faster than ever before) are security risks for everybody. President Bush declared at the 1992 Earth Summit that American life-styles were not up for negotiation on ways to change. Indeed not. They have already committed Americans to a track that will bring changes as momentous as they are inevitable. That much has been true ever since Americans embarked on a gasoline binge forty years ago, ever since they started to chop down their forests at a rate faster than those of Brazilian Amazonia, and ever since they caused the soil of Iowa to erode as fast as that of Ethiopia.

Not all Americans are like that. Far from it. One in fifteen belongs to an environmental organization, and at recent rates of membership increase the figure will soon rise to one in ten.

There is a long way before they match the record of Denmark, where there are more environmentalists, as registered through membership, than there are Danes. But when American environmentalists voice their views through their votes, the politicians listen; and when they also vote with their dollars in the marketplace, business leaders follow their lead.

And yet, and yet. Twice a year I head for the United States (last spring was my seventieth visit) in order to work the lecture circuit. On a recent occasion I made my preparations in Oxford, England, and they included buying a set of shirts. I asked for an opinion of them from my two late-teen daughters. "Fine, Dad," they said; "they'll look good on you. By the way, where were they made?" I looked inside the collar: "Made in Britain." Another small act that helps drive some clothing worker in Brazil onto the streets and perhaps into the rainforests or some other lands that are environmentally marginal for human use. Touché.

Think too about the age we live in and the remarkable changes that demand a radical rethinking from every one of us. The following examples show that we are in the midst of a whirlwind of change even though we often continue on our way as if we feel scarcely a breeze on our faces.[2] The sooner we can wrap our minds around the following, the sooner we shall be ready to comprehend the seismic shifts in four major factors of our times. These four are the ways of nations, the makeup of the global community, the support role of our environments, and the changing nature of our security.

• During the 1980s we added more than $4.5 trillion to the global economy, an amount that exceeded the entire economy in 1950. One-quarter of this additional sum was spent on military activities.

• We have pushed ahead with the opening phase of what promises to be the biggest mass extinction of species in 65 million years.

· There are now 2 billion radios in the world, twice as many as in 1975. News in one part of the world becomes same-day news everywhere on Earth. The community of humankind is finally becoming a single community.

· A few leading multinational corporations have now achieved such commercial clout that they generate annual sales worth more than the entire economies of all but a handful of countries. A single corporation can mold our collective future as much as one hundred governments. If used for constructive purposes, this global reach is one of the most powerful unifying forces ever to emerge. Regrettably, that is not often the case to date.

· For the first time in human history, we have engaged in a shift, to be played out within a few human generations, from being a reproductive to a contraceptive society in every part of the world.

· Whereas the 1980s were the "me decade," the 1990s will be the "we decade." Either this will occur because we want it and we gain it, all of us winners together, or it will occur because we pretend it can wait, whereupon we shall all—all of us together again—end up losers. This makes us a "we" generation beyond all previous experience.

Profound changes. We are all part of them, we all contribute to them, and we shall all enjoy or regret their consequences. We can all help to shape their outcome. Every last one of us plays his or her part, whether positive or negative, whether deliberate or regardless. We are witnessing the arrival of *Homo sapiens* as the most dominant species to have existed, enveloping the whole Earth habitat with its activities. We are becoming humankind—also human kind?

Above all, let us bear in mind that this is a time, more than any time in the past thanks to telecommunications and other forms of participation in the world around us, when everybody can be somebody and nobody need be anybody. Let us remember, too, the dictum of that redoubtable American Margaret Mead: "Never doubt that a small group of thoughtful committed citizens can change the world. Indeed it's the only thing that ever has."

What We Can All Do

What can the individual citizen do to contribute to the cause? To make the question more direct and specific, what can you, the reader, do to support the global campaign to safeguard our global habitat? Here is a short list of practical items that you can undertake tomorrow morning—or better, this evening.

1. Support the environmental organization of your choice. Whether it is the World Wildlife Fund or Friends of the Earth or any other of a dozen leading bodies, write that subscription check. If you already belong to an organization, join another. The cost is generally less than that of a cocktail a month. All of them squeeze superlative value out of every last dollar, and all are oriented, whether deliberately or not, toward environmental security.

2. Write another piece of paper that costs far less but seems to cost more insofar as not all environmentalists get around to doing it. Send off a letter to your senator or representative, copied to the president, urging more action on an environmental issue. Do it again in a couple of weeks' time, focusing on a different environmental question. Cost: ten minutes of your time, helping to save the planet from injuries that could persist for thousands of years.

During the course of a two-hour breakfast with then-Senator Albert Gore (foreigners need extra time to climb their way through a pile of American waffles), I asked him why his many bills for environmental legislation usually failed to muster enough votes among his senatorial colleagues to pass into law.

Don't they believe in it? Of course they do, he replied. Trouble was they did not think there were enough environmentalist votes back home to warrant the time they would have to spend on getting themselves up to speed on the nitty-gritty of Gore's proposals. If only, he mused, environmentalists would get those letters flowing, efforts on the Senate floor would be vastly more successful. I could not help recalling that the National Rifle Association, with only one-fifth as many members as environmentalist groups, can exert critical leverage over legislation by the simple expedient of getting half a million letters onto Capitol Hill within days of the action call going out to members. That is a measure of what can be achieved by committed citizens. Don't environmentalists care as much for their cause?

These two items are obvious enough, little though they are practiced by certain would-be environmentalists. Let's now move on to some less apparent initiatives.

3. Cut back on excessive or merely wasteful consumerism, recycle like crazy, and generally tread more lightly on the face of the Earth. Taste the satisfaction of doing more with less at dozens of points in your daily round. Watch out for your quality of life rather than quantity of livelihood (remember that nobody says on their deathbed that they wish they had made more money).

For some specific examples of what we can all do to cut back on global warming by burning less fossil fuel, consider that an American family can reduce its carbon emissions by a whopping two tons a year. All it has to do is this. Gain five miles per gallon of gasoline through trading in a gasoline guzzler for a more efficient car—an initiative worth 2000 pounds of carbon per year. Keep the car's engine tuned, worth another 900 pounds. Drive less, for example, by combining errands, worth 500 pounds. Keep your car tires properly inflated, 250 pounds. Recycle one aluminum can, one glass bottle, and one newspaper each day, 290 pounds. Replace a conventional 100-watt bulb with a 27-watt compact fluorescent bulb, 160 pounds.

Plant a tree, 150 pounds. Turn off unnecessary lights, 120 pounds. For further details, contact the Rocky Mountain Institute in Snowmass, Colorado, or the World Wildlife Fund–U.S. in Washington, D.C.

4. Ask yourself if your life-style causes you to engage in activities that inadvertently harm environments around the back of the world. For myself, I constantly remind myself of the "new shirt" syndrome described above. By virtue of the global economy, we are all involved in everybody else's welfare. Ask yourself how your daily purchases and related activities affect people in Brazil, Kenya, and India—especially the impoverished people who, for want of employment, are likely to pick up their machetes and matchboxes and head off to tropical forests.

5. Keep an eye on the biggest ball of all. We live at a major watershed in the entire course of the human enterprise. During the next few decades at most, we shall determine the future security of our planet for thousands if not millions of years. Either we shall save it for more generations than since humans first became humans, or we shall consign them to an Earth habitat ravaged in ways unprecedented in the entire course of the human adventure. No doubt about it, we live at a privileged time. It is in our hands, and our hands alone. Shouldn't we count ourselves fortunate beyond dreams, that we have the chance to save the planet at a time of unparalleled threat? What a superb challenge! Who is there who cannot wait to get on with the job?

This is no more than a preliminary list of readily available actions. A still further item is to expand the list with your own ideas until it becomes more of your personal list. Use your imagination, be creative, and come up with a whole line of items geared to your particular preferences. Above all, and to cite another pop song, "Why don't you do it? Everybody's doing it. Get on and do it, do it, do it."

The Future and
Those with the Biggest Stake

During the past few years I have had the pleasure of short teaching stints at universities on both sides of the Atlantic. I delight in the company of students. I prefer to schedule classes in late afternoon, so that afterwards we can all adjourn to a locale for pizzas, beer, and more discussion. I have been fortunate in the students. They have taught me a lot. What I have learned is this.

They have the biggest stake in the future because they have longest to live there, even though they have little say in how it shall turn out. (They also have to do the fighting whenever their elders and betters decree that the best way to run the world is through violence.) Many of them sense little confidence in the future. They feel that a worthwhile future is being denied them even before they are old enough to do much about it. "Trust nobody over thirty." Some of them fear their future will be filled with gunsmoke.

Yet they retain a sense that there is still much to be won. Widely traveled as they often are, they too see a world without the straitjacket of political boundaries. It is an experience that some of them could have shared with their great-grandparents. In Europe before 1914, people could roam from one country to another without passports. So, too, today: the great experiment in social engineering that has become the European Community no longer requires passports for travel from France to Germany, from Ireland to Greece. Whatever the past value of national boundaries, enlightened politicians have determined that Community members are better off without them. In light of this pioneering example, could the next generation find that extremist forms of nationalism in other parts of the world will shortly be consigned to history? And will they be looked back upon as a brief aberration in the human enterprise?

So while we are conducting a planetwide experiment with our environments, we are also conducting a number of mini-experiments on ways to run our world.

Still another phenomenon of our times is surely an aberration. It is our reluctance to look ahead to our longer-term future. Many of our environmental problems stem from a curious kind of myopia. When bankers impose an interest rate of 10 percent, they are effectively saying (because of discounting) that there is no future beyond seven years. Politicians tell us by their actions (and inaction) that it is absurd for them to consider the future beyond the next election. Business leaders display even less of a futurist inclination; the star they navigate by is the bottom line of the next quarter. A dozen times an hour we are assured it is smart and even sophisticated to live for the day and the goodies thereof.

It has not always been this way. Almost invariably it has been quite different. At dozens of places in Europe, the Gothic cathedrals are witness to societies and cultures that would invest immeasurably in the future. The planners and builders of the cathedrals usually knew they would not live to see the finished product. The same with the people who toiled to construct the Pyramids and the Great Wall of China. In virtually all ages except our own, people lived with a keener eye on the future. Only we of the late twentieth century are preoccupied with the consumerist present to the grand-scale detriment of what we are doing to the future—as for hundreds of our descendants' generations.

True, we make a massive investment in our children. What a splendid gesture of faith in the future, to bring a new life into the world. Yet through our fixation on now, we are undercutting the very foundations of our children's future. It costs about $200,000 to bring up a child from cradle through college. With a comparatively small outlay to reinforce our original investment, we could assure a world with air fit for our children to breathe. But a hundred times a day we are exhorted to take no thought for the morrow.

So when I am enjoying my privileged days on campus, I tell the students that the times are out of joint. To restore them will be a formidable task. In fact, there are occasions when I myself feel more than daunted by the prospect. I sometimes wake up in the morning to face the problems out there, and I am reminded of the person who defined an optimist and a pessimist. An optimist, he said, proclaims this is the best of all possible worlds; to which the pessimist responds that it is probably true. But then I think of those students and their steadfast idealism, their commitment to the prospect of building that shining citadel on the hilltop.

Some Better News

To hearten us on our way, let us remind ourselves of some "good news" items. There are quite a few, and we certainly need all we can find. True, they are isolated incidents and all too rare, and taken together they make up only a fraction of what we need. But they are there, they are bright stars in an otherwise dark sky, and they prove to us that when we bestir ourselves we can achieve remarkable breakthroughs. How much better than ten years ago, when the sky was almost horizon-to-horizon black. Here's a sampler list of "good news" items—and by way of planetary-citizen homework, seek out more items.

1. We now save 2.5 million developing-world children a year who would otherwise die of malnutrition or disease. While there are still another 13 million children to save, we are doing much better than a few years back.

2. A number of American communities have decided they will not wait for the U.S. government to get its act together on global warming. Vermont and Oregon, soon to be followed by California, Maryland, and New York, have chosen to go it alone by reducing their carbon dioxide emissions by the year 2000 far more than the U.S. government aims to do. Similarly, a Connecticut utility company has planned to plant enough trees in Guatemala to offset its carbon emissions back home. These are examples of social responsibility at a global level.

3. Los Angeles, the land where the car has been king, has decided that the gasoline engine shall be phased out by the year 2000.

4. In Haiti, likely the world's worst environmental basket case, 200,000 farmers have recently planted 50 million trees through the support of citizen-activist groups both within the country and from the United States.

5. A Manhattan couple, Glen and Millie Leat, show what can be done at micro-scale through their Trickle Up program. They make loans of no more than $100 a time to developing-country entrepreneurs. Total loans to date, 12,000 in 83 countries; as many as 100,000 are expected by the year 2000. Payback rate, 96 percent—an achievement much better than the World Bank manages.

6. Thailand's extraordinary achievement in bringing its family size down from over six children in 1970 to little more than two today has been due primarily to the initiatives of a family-planning activist, Mechai Viravaidya. When you visit Bangkok, you see billboards all over town proclaiming the virtues of the two-child family. You hear the message hourly on the radio, you see it in advertisements in the cinema, you hear it in many a local pop song, you encounter the spiel dozens of times a day—and if you go to see national sporting teams, you read it emblazoned on their shirts.

7. Costa Rica has maintained its record of managing without any military forces whatever, even while the other Central American countries have been racked by one violent upheaval after another. Costa Rica also has more teachers than police.

8. Zimbabwe has shown how to shift from a major food importer into a sizable food exporter. Except for the recent drought years, it has proved throughout the 1980s to be one of the few sub-Saharan African countries that could learn to feed itself again and find a surplus to send to famine-stricken countries such as Ethiopia.

9. The Colombian government long allowed its sector of Amazonia to be burned for cattle ranches and small-scale farmers. In 1989, it decided to assign almost half of its remaining forest expanse to the care of tribal peoples, who have demonstrated for centuries that they know how to gain livelihoods from the forest without knocking it over. Similarly, Brazil has rescinded many of its government subsidies for Amazonia ranches, and Bolivia has declared a moratorium on logging. Both India and Vietnam have launched forestry strategies with emphasis on conservation as well as exploitation.

10. Most readers of this book will remember when the world was subject to the scourge of smallpox. By 1980 the disease was finally eliminated. Cost of the decades-long effort, $350 million; benefits, at least $1 billion a year. As noted above, we could readily do the same for malaria, plus several other major killer diseases. The payoff in economic terms alone would be greater than we could generally find on Wall Street.

The Great Creative Challenge

I write this concluding chapter during the summer of 1992 while waiting for the Barcelona Olympics to begin. I have been a runner all my life, and now that I am in my late fifties I find I am still fit enough to leave thousands of men in their twenties behind in mass marathons. So I revel in international athletics, especially the Olympics. This is youth doing its thing, and then some. But I regret the raw nationalism that overtakes the Olympics, with country-by-country medal tables. The partici-

pants generally do not care much. For them it is a case of
Johnny Gregson versus Maina Kiprotich versus Mahendra
Singh, rather than America versus Kenya versus India. I suspect many of them share my dream that one day we shall see
all competitors wearing a plain white vest. Then we shall all
have taken another step in our joint scaling of other Olympian
heights.

To do that we shall have to be tough, deploying all the
mental muscle we can muster. We shall be taking on a challenge that has proved to be beyond the capacities of humanity
to date. We have had plenty of seers for centuries, pointing the
way to the promised land of global unity. We have steadfastly
ignored them, or simply felt we were not up to the trip, or tried
out alternatives. Now we have only one choice. It has to be
taken sooner or later, and to be taken via a golden-age outlook
or a heaven-forbid scenario.

The trip will not be easy. I know this from my own experience. Much as I proclaim the virtues of a planetary mind-set, I
still find myself subject to ancient atavisms. In my youth in the
north of England, I used to go to cheer the local football club,
Blackburn Rovers. When I left Britain to live in Kenya, I remained a Rovers supporter in strong spirit. On Saturday evening, nothing would do but I must buy the latest-edition newspaper in Nairobi to find out the afternoon's result from the
great team. I used to wonder why I remained so involved with
the Rovers. Most of the time I could not name more than a
couple of players. Yet the pull was there. An attachment from
childhood that I should leave behind? Surely not; no harm in
keeping up a link of that sort. But it bespeaks an instinctive
harking-back that, played out in the larger arena of societies,
nations, and governments, is no longer a risk-free activity.

All of us sense such loyalties. First to the family. Then the
local community, complete with football club or whatever.
Next the county, then the state, then the country. Thereafter
the region, the continent—and hopefully the world? Loyalty to

one level does not mean any less loyalty to other levels. Yet we seem to back off from that last stride toward global citizenship. We already identify on several levels: why not the widest level, too, especially when it is now so clearly for our own good as well as the world's?

The challenge, then, is not one of knowledge or intellectual understanding. Still less is it a case of science or technology. Least of all is it a question of money when many Northerners are as rich as Croesus or Louis XIV. The key ingredient is surely imagination—that creativity of the emotional and even spiritual side of ourselves that should spur us to reach out and be good neighbors to all our neighbors, all 5.5 billion of them. Can we climb onto our global citizen's feet and stand tall enough to see beyond the restricted and outmoded horizons of tradition? Can we recognize that our shortsighted gaze may cause those horizons to become polluted beyond all previous experience—polluted either with industrial smoke or with the smoke of gunfire?

The Ultimate Payoff

To wind up this book, what better way than to quote the words of that great environmental president, Teddy Roosevelt. He accepted that people of limited vision will sometimes feel affronted by those who look further. But:

> It is not the critic who counts, not the man who points out how the strong man stumbled or how the doer of deeds could have done better. The credit belongs to the man who is actually in the arena, whose face is marred by dust and sweat and blood, who strives valiantly, who errs and comes short again and again, who knows the great enthusiasm, the great venture, and spends himself in a worthy cause, who at the best knows in the end the tri-

umph of high achievement, and who at the worst, if he
fails, at least fails while daring greatly so that his place
shall never be with those cold and timid souls who know
neither victory nor defeat .

[T. Roosevelt, *A Strenuous Life*,
Grant Richards, London, 1906].

Notes

CHAPTER 1: What It's All About

1. World Commission on Environment and Development, *Our Common Future* (New York: Oxford University Press, 1987).

2. N. Myers, "The Environmental Dimension to Security Issues," *The Environmentalist,* 6(1986):251–257; N. Myers, "Population, Environment and Conflict," *Environmental Conservation,* 14(1987):15–22; N. Myers, "Environment and Security," *Foreign Policy,* 74(1989):23–41; N. Myers, "Population Growth, Environmental Decline and Security Issues in SubSaharan Africa," in A. Hjort af Ornas and M. A. Mohamed Salih, eds., *Ecology and Politics: Environmental Stress and Security in Africa* (Uppsala, Sweden: Scandinavian Institute of African Studies, 1989), pp. 211–231; N. Myers, "Environmental Security: The Case of South Asia," *International Environmental Affairs,* 1(1989):138–154.

3. P. R. Ehrlich and A. H. Ehrlich, *The Environmental Dimensions of National Security* (Stanford, CA: Stanford Institute for Population and Resource Studies, 1988); J. Galtung, *Environment, Development and Military Activities* (New York: Columbia University Press, 1982); P. H. Gleick, *Environment, Resources, and Security: Arenas for Conflict, Areas for Cooperation* (Berkeley, CA: Pacific Institute for Studies in Development, Environment and Security, 1990); P. H. Gleick, "Environment and Security: The Clear Connections," *Bulletin of the Atomic Scientists,* 47(1991):17–21 [for a critical response, see D. Deudney, "Environment and Security: Muddled Thinking," *Bulletin of the Atomic Scientists,* 47(1991):22–28]; T. F. Homer-Dixon, "On

the Threshold: Environmental Changes as Causes of Acute Conflict," *International Security,* 16(1991):76–116; R. D. Lipschutz and J. P. Holdren, "Crossing Borders: Resource Flows, the Global Environment and International Security," *Bulletin of Peace Proposals,* 21(1990):121–133; J. T. Mathews, "Redefining Security," *Foreign Affairs,* 68(2, 1989):162–177; Myers, "Environment and Security"; G. Porter, "Post–Cold War Global Environment and Security," *Fletcher Forum of World Affairs,* 14(2, 1990):332–344; J. Tinker and L. Timberlake, *Environment and Conflict* (London: Earthscan Publications, 1985); T. F. Homer-Dixon, J. H. Boutwell, and G. W. Rathjens, "Environmental Change and Violent Conflict," *Scientific American,* 268(2, 1993):38–45; C. Thomas, *The Environment in International Relations* (London: Royal Institute of International Affairs, 1992); R. H. Ullman, "Redefining Security," *International Security,* 8(1983):129–153; A. H. Westing, ed., *Global Resources and International Conflict* (New York: Oxford University Press, 1986).

4. M. S. Gorbachev, At the General Assembly of the United Nations, New York, December 1988.

CHAPTER 2: Environmental Security: How It Works

1. This is not to overlook another environmental dimension to security concerns, in the form of damage done to the environment through, for example, Americans' defoliation in Vietnam or Saddam Hussein's torching of Kuwaiti oil wells [see A. Ehrlich and J. W. Birks, eds., *Hidden Dangers: Environmental Costs of Preparing for War* (San Francisco: Sierra Club Books, 1990)]. But this book concentrates on environmental problems as causes rather than consequences of conflict.

2. J. Goldstone, *Revolution and Rebellion in the Early Modern World* (Berkeley, CA: University of California Press, 1990); R. D. Lipschutz, *When Nations Clash: Raw Materials, Ideology, and Foreign Policy* (New York: Ballinger, 1989); C. Ponting, *A Green History of the World* (London: Sinclair-Stevenson, 1991); G. Porter, "Post–Cold War Global Environment and Security," *Fletcher Forum of World Affairs,* 14(2, 1990):332–344; W. Youngquist, *Mineral Resources and the Destinies of Nations* (New York: National Book Company, 1990).

3. For background reading on this central issue, see the items listed under Chapter 1, Note 3.

4. J. W. Sewell and S. K. Tucker, eds., *Growth Exports and Jobs in a Changing World Economy* (New Brunswick, NJ: TransAction Books, 1988); J. W. Suomela, *The Effects of Developing Countries' Debt Servicing Problems on U. S. Trade* (Washington, D.C.: International Trade Commission, 1987).

5. Suomela, *Developing Countries' Debt Servicing Problems;* Sewell and Tucker, *Growth Exports and Jobs;* S. K. Tucker, *The Debt-Trade Linkages to*

U.S.–Latin American Trade (Washington, D.C.: Overseas Development Council, 1990).

6. F. C. Ikle and A. Wohlstetter, *Discriminate Deterrence: Report for the Department of Defense of the Commission on Integrated Long-Term Strategy* (Washington, D.C.: U.S. Government Printing Office, 1988).

7. L. R. Brown, *Redefining National Security* (Washington, D.C.: Worldwatch Institute, 1977).

8. R. S. McNamara, *The Essence of Security;* cited in *End to Hunger* (New York: Praeger, 1968).

CHAPTER 3: The Middle East

1. M. Falkenmark, "Middle East Hydropolitics: Water Scarcity and Conflicts in the Middle East," *Ambio,* 18(1989):350–352; T. F. Homer-Dixon, J. H. Boutwell, and G. W. Rathjens, "Environmental Change and Violent Conflict," *Scientific American,* 268(2, 1993):38–45; T. Naff, ed., *Water Issues in the Middle East* (Philadelphia: Associates for Middle East Research, 1990); T. Naff, *The Jordan River System: Political, Socioeconomic, and Strategic Issues: Jordan and Israel* (Philadelphia: Associates for Middle East Research, 1990); E. Salameh, *Water and Other Resources in the Middle East, and Their Developmental and Strategic Importance up to 2002* (Amman, Jordan: Department of Hydrology and Engineering, University of Jordan, 1989); J. R. Starr, "Water Wars," *Foreign Policy,* 82(1991):17–36.

2. E. Salameh, *Water Resources of the Jordan River System and the Surrounding Countries: Significance and Implications for Socioeconomic Development* (Amman, Jordan: Department of Hydrology and Engineering, University of Jordan, 1989).

3. J. R. Starr and D.C. Stoll, eds., *The Politics of Scarcity: Water in the Middle East* (Boulder, CO: Westview Press, 1988).

4. A. Soffer and N. Kliot, *Regional Water Projects in the Middle East* (Haifa, Israel: University of Haifa, 1988); D. M. Wishart, "The Breakdown of the Johnstone Negotiations Over the Jordan Waters," *Middle Eastern Studies,* 26(1990):536–546; Salameh, *Water Resources of the Jordan River System.*

5. I. P. Beaumont and K. McLaughlin, eds., *Agricultural Development in the Middle East* (New York: John Wiley, 1985); M. Evenari, "The Problem Posed: Elements of the Agriculture Crisis," *Israel Journal of Development,* 10(1988):4–8; H. I. Shuval, ed., *Water Quality Management Under Conditions of Scarcity: Israel as a Case Study* (New York: Academic Press, 1980).

6. A. M. Farid and H. Sirriyeh, eds., *Israel and Arab Water* (London: Ithaca Press, 1985); Naff, *Water Issues.*

7. N. Choucri, *Challenges to Security: Population and Political Economy in*

Egypt (London: Harper Collins Academic, 1992); Economist Intelligence Unit, *Egypt Country Profile 1986–87* (London: The Economist Publications, 1988); Food and Agriculture Organization, *Potential Population Supporting Capacities of Lands in the Developing World* (Rome: Food and Agriculture Organization, 1984); P. H. Gleick, "The Vulnerability of Runoff in the Nile Basin to Climatic Changes," *The Environmental Professional,* 13(1991):66–73; Starr and Stoll, *The Politics of Scarcity.*

8. Salameh, *Water and Other Resources in the Middle East.*

9. H. J. Shuval, "The Development of Water Re-Use in Israel," *Ambio,* 16(1987):186–190; Starr and Stoll, *The Politics of Scarcity.*

10. Starr, "Water Wars."

11. Starr and Stoll, *The Politics of Scarcity.*

12. Starr, "Water Wars."

13. Choucri, *Challenges to Security;* P. H. Gleick, "Climate Changes, International Rivers, and International Security: The Nile and the Colorado," in R. Redford and T. J. Minger, eds., *Greenhouse Glasnost: The Crisis of Global Warming* (New York: Ecco Press, 1990), pp. 147–165; Gleick, "The Vulnerability of Runoff"; P. H. Gleick, ed., *A Guide to the World's Water Resources: The Coming Crisis* (Berkeley, CA: Pacific Institute for Studies in Development, Environment and Security, 1992).

14. Choucri, *Challenges to Security;* P. Jabber, "Egypt's Crisis: America's Dilemma," *Foreign Affairs,* 64(5, 1986):960–980.

15. World Bank, *World Development Report 1992* (New York: Oxford University Press, 1992).

16. M. A. Kishk, "Land Degradation in the Nile Valley," *Ambio,* 15(1986):226–230.

17. E. D. Stains, *Irrigation Briefing Paper* (Cairo: Office of Irrigation and Land Development, U.S. Agency for International Development, 1987).

18. R. S. Bradley, H. F. Diaz, J. K. Eischard et al., "Precipitation Fluctuations Over Northern Hemisphere Land Areas Since the Mid-19th Century," *Science,* 237(1987):171–175; Gleick, "The Vulnerability of Runoff"; J. A. Mabbutt, "Impacts of Carbon Dioxide Warming on Climate and Man in the Semi-Arid Tropics," *Climatic Change,* 15(1989):191–221.

19. M. O. Beshir, ed., *The Nile Valley Countries: Continuity and Change* (Khartoum: Institute of African and Asian Studies, University of Khartoum, 1984); B. A. Godana, *Africa's Shared Water Resources: Legal and Institutional Aspects of the Nile, Niger and Senegal River Systems* (Boulder, CO: Lynne Rienner Publications, 1985); D. Jovanovic, "Ethiopian Interests in the Division of the Nile River Waters," *Water International,* 10(1985):82–85; Starr and Stoll, *The Politics of Scarcity;* J. Waterbury, "Legal and Institutional Arrangements for Managing Water Resources in the Nile Basin," *Water Resources Development,* 3(1987):92–103; D. Whittington and K. E. Haynes, "Nile Water for Whom? Emerging Conflicts in Water Allocation for

Agricultural Expansion in Egypt and Sudan," in P. Beaumont and K. McLaughlin, eds., *Agricultural Development in the Middle East* (New York: John Wiley, 1985).

20. A. Gore, *Earth in the Balance* (Boston: Houghton Mifflin, 1992).

21. A.-A. Cashed, "The Nile—One River and Nine Countries," *Journal of Hydrology*, 53(1981):53–84; Jovanovic, "Ethiopian Interests"; M. Shahin, *Hydrology of the Nile Basin* (New York: Elsevier, 1985).

22. A. Sadat, cited in "Egypt: Threat to Nile Water," *African Recorder*, 19(1985):5, 396.

23. B. Ghali, cited in "Egypt: Threat to Nile Water."

24. Falkenmark, "Middle East Hydropolitics"; Gleick, "Climate Changes"; Gleick, *A Guide to the World's Water Resources*; Naff, *Water Issues*; Salameh, *Water Resources of the Jordan River System*; Starr, "Water Wars"; Starr and Stoll, *The Politics of Scarcity*.

25. P. H. Gleick, "The Implications of Global Climatic Changes for International Security," *Climatic Change*, 15(1989):309–325.

26. Ibid.

27. M. Falkenmark, "Rapid Population Growth and Water Scarcity: The Predicament of Tomorrow's Africa," in K. Davis and M. D. Bernstam, eds., *Resources, Environment and Population* (New York: Oxford University Press, 1991), pp. 81–94; M. Falkenmark, J. Lundqvist, and C. Widstrand, *Water Scarcity—An Ultimate Constraint in Third World Development* (Linkoping, Sweden: Department of Water and Environmental Studies, University of Linkoping, 1990).

28. Ibid.

29. Falkenmark, "Middle East Hydropolitics"; M. J. W. la Riviere, "Threats to the World's Water," *Scientific American*, 261(3, 1989):80–94; L. Nelson and C. Sandell, *Population and Water Resources* (New York: National Audubon Society, 1991).

30. S. Postel, *Water for Agriculture: Facing the Limits* (Washington, D.C.: Worldwatch Institute, 1989).

31. World Health Organization, *Review of Progress of the International Drinking Water Supply and Sanitation Decade, 1981–1990* (Geneva: World Health Organization, 1988). See also M. Munasinghe, *Managing Water Resources to Avoid Environmental Degradation: Policy Analysis and Application* (Washington, D.C.: Environment Department, The World Bank, 1990); World Bank, *Water and Sanitation: Toward Equitable and Sustainable Development* (Washington, D.C.: The World Bank, 1988).

32. World Commission on Environment and Development, *Our Common Future* (New York: Oxford University Press, 1987).

33. R. Clarke, *Water: The International Crisis* (London: Earthscan Publications, 1991); Gleick, *A Guide to the World's Water Resources*; la Riviere, "Threats to the World's Water."

34. L. R. Brown et al., *State of the World 1989* (New York: W. W. Norton, 1989).

35. Starr, "Water Wars."

CHAPTER 4: Ethiopia

1. D. W. Georgis, *Red Tears: War, Famine and Revolution in Ethiopia* (Trenton, NJ: Red Sea Press, 1989); F. G. Kiros, "Economic Consequences of Drought Crop Failure and Famine in Ethiopia, 1973–1986," *Ambio*, 20(1991):183–185; S. Rubenson, "Environmental Stress and Conflict in Ethiopian History: Looking for Correlations," *Ambio*, 20(1991):179–182.

2. N. Gebramendhin, *The Environmental Dimension of Security in the Horn of Africa: The Case of Somalia* (Nairobi: United Nations Environment Programme, 1991).

3. M. Chege, "Conflict in the Horn of Africa," in E. Hansen, ed., *Africa: Perspectives on Peace and Development* (London: Zed Books, 1987), 87–100; Georgis, *Red Tears;* G. Hancock, *Ethiopia: The Challenge of Hunger* (London: Gollancz Publishers, 1985); J. Markakis, *Conflict in the Horn of Africa* (New York: Cambridge University Press, 1987); R. K. Molvaer, *Environmental Security in the Horn of Africa: An Annotated Bibliography* (Nairobi: United Nations Environment Programme, 1989); B. D. Perry, "The Real Cause of Ethiopia's Problems," *Nature*, 319(1986):183; S. P. Petrides, *The Boundary Question between Ethiopia and Somalia* (New Delhi: New Social Press, 1983).

4. Gebramendhin, *The Case of Somalia.*

5. T. J. Farer, *War Clouds on the Horn of Africa* (Washington, D.C.: Carnegie Endowment for International Peace, 1979); J. Shepherd, *The Politics of Starvation* (Washington, D.C.: Carnegie Endowment for International Peace, 1975).

6. B. H. Selassie, "The American Dilemma on the Horn of Africa," *Journal of Modern African Studies*, 22(1984):257–264.

7. D. A. Korn, *Ethiopia, The United States and the Soviet Union* (London: Croom Helm, 1986); R. Luckham and D. Bekele, "Foreign Powers and Militarism in the Horn of Africa," *Review of African Political Economy*, 30(1984):8–20, 31(1984):7–28; M. Ottaway, *Soviet and American Influence in the Horn of Africa* (New York: Praeger, 1982); P. Woodward, *Rivalry and Conflict in North-East Africa* (London: Centre for Security and Conflict Studies, Institute for the Study of Conflict, 1987).

8. G. Kebbede and M. J. Jacob, "Drought, Famine and the Political Economy of Environmental Degradation in Ethiopia," *Geography*, 73(1988):127–143; J. Markakis, *National and Class Conflict in the Horn of Africa* (London: Zed Books, 1990); W. B. Selassie, *Conflict and Intervention in the Horn of Africa* (New York: Monthly Review Press, 1980); R. Ulrich, *Environment*

and Security in the Horn of Africa (Nairobi: United Nations Environment Programme, 1989).

9. K. Newcombe, *An Economic Justification for Rural Afforestation: The Case of Ethiopia* (Washington, D.C.: The World Bank, 1984).

10. D. W. Pearce, *The Economics of Natural Resource Management: Issues Paper* (Washington, D.C.: The World Bank, 1986).

11. P. Cutler, "The Political Economy of Famine in Ethiopia and Sudan," *Ambio*, 20(1991):176–178; K. Jansson, M. Harris, and A. Penrose, *The Ethiopian Famine* (Atlantic Highlands, NJ: Humanities Press International, 1987); A. Westing, "Environmental Security and Its Relation to Ethiopia and Sudan," *Ambio*, 20(1991):168–171.

12. T. Kidane-Mariam, *Ethiopia: An Overview of Its Priority Environmental Problems, Policy Directives, Objectives, Strategies and Targets* (Nairobi: United Nations Environment Programme, 1985); M. Wold-Marian, "An Assessment of Stress and Strain on the Ethiopian Highlands," *Mountain Research and Development*, 8(1988):259–264.

13. World Bank, *World Development Report 1991* (Washington, D.C.: The World Bank, 1991).

14. H. Hurni, *Towards Sustainable Development in Ethiopia* (Bern: The Geography Institute, Bern University, 1990); see also R. Barber, *An Assessment of the Dominant Soil Degradation Processes in the Ethiopian Highlands* (Addis Ababa: Ethiopian Highlands Reclamation Study, Ministry of Agriculture, 1984); K. Tato and H. Hurni, eds., *Soil Conservation for Survival* (Bern: The Geography Institute, Bern University, 1991).

15. D. W. Pearce, ed., *Blueprint 2: Greening the World Economy* (London: Earthscan Publications, 1991).

16. R. A. Hutchison, ed., *Fighting for Survival: Insecurity, People and the Environment in the Horn of Africa* (Gland, Switzerland: International Union for Conservation of Nature and Natural Resources, 1991); World Bank, *Ethiopia—Food Security Study* (Washington, D.C.: Agricultural Operations Division, The World Bank, 1991).

17. R. Chambers, "Hidden Losers? The Impact of Rural Refugees and Refugee Programs on Poorer Hosts," *International Migration Review*, 2(1986):254–263; R. Ek and A. Karadawi, "Implications of Refugee Flows on Political Stability in the Sudan," *Ambio*, 20(1991):196–203; J. R. Rogge, *Too Many, Too Long: Sudan's Twenty-Year Refugee Dilemma* (Totowa, NJ: Rowan and Allanheld, 1985); Westing, "Environmental Security."

18. D. Gamachu, *Environment and Development in Ethiopia* (Geneva: International Institute for Relief and Development, 1988); Hutchison, *Fighting for Survival;* F. G. Kiros, "Economic Consequences of Drought Crop Failure and Famine in Ethiopia 1973–1986," *Ambio*, 20(5, 1991):183–185; S. Rubenson, "Environmental Stress and Conflict in Ethiopian History: Looking for Correlations," *Ambio*, 20(1991):179–182.

19. A. H. Ehrlich, "Critical Masses," *The Humanist,* 45(1985):18–22, 36.

20. Hutchison, *Fighting for Survival.*

21. A. R. A. Z. Ahmed et al., *War Wounds: Development Costs of Conflict in Southern Sudan* (London: The Panos Institute, 1988); A. A. Ali, *Sudan Economy in Disarray* (London: Ithaca Press, 1987); N. Cater, *Sudan: The Roots of Famine* (Oxford: Oxfam Publications, 1986); P. Cutler, "The Political Economy of Famine in Ethiopia and Sudan," *Ambio,* 20(1991):176–178.

22. N. Myers, *The Wild Supermarket* (Gland, Switzerland: World Wildlife Fund International, 1986).

CHAPTER 5: Sub-Saharan Africa

1. B. T. Chidzero, "Africa and the World Economy," *World Futures,* 25(1988):157–162; J. P. DeCuellar, *Africa's Economic Situation* (New York: Office of the Secretary General, United Nations, 1988); World Bank, *World Development Report 1992* (Washington, D.C.: The World Bank, 1992).

2. A. M. Diallo, *Population Policies and Family Planning in Sub-Saharan Africa* (Harare, Zimbabwe: All-Africa Council on Population and Development, 1990); L. Huss-Ashmore and S. H. Katz, eds., *African Food Systems in Crisis* (New York: Gordon and Breach, 1989); R. S. McNamara, "Population and Africa's Development Crisis," *Populi,* 17(4, 1990):35–43.

3. World Bank, *The Challenge of Hunger in Africa: A Call to Action* (Washington, D.C.: The World Bank, 1988); World Bank, *World Development Report 1991* (Washington, D.C.: The World Bank, 1991).

4. Food and Agriculture Organization, *State of Food and Agriculture 1991* (Rome: Food and Agriculture Organization, 1991).

5. R. S. Bradley, H. F. Diaz, J. K. Eischeid et al., "Precipitation Fluctuations Over Northern Hemisphere Land Areas Since the Mid-19th Century," *Science,* 237(1987):171–175; J. A. Mabbutt, "Impacts of Carbon Dioxide Warming on Climate and Man in the Semi-Arid Tropics," *Climatic Change,* 15(1989):191–221; see also World Bank, *World Development Report 1992.*

6. McNamara, "Population and Africa's Development Crisis."

7. L. R. Brown and E. C. Wolf, *Reversing Africa's Decline* (Washington, D.C.: Worldwatch Institute, 1985); T. J. Goliber, *Sub-Saharan Africa: Population Pressures on Development* (Washington, D.C.: Population Reference Bureau, 1985; N. Myers, "Population Growth, Environmental Decline and Security Issues in Sub-Saharan Africa," in A. Hjort af Ornas and M. A. M. Salih, eds., *Ecology and Politics: Environmental Stress and Security in Africa* (Uppsala, Sweden: Scandinavian Institute of African Studies, 1989), pp. 211–229; F. T. Sai, "Population Factor in Africa's Development Dilemma," *Science,* 226(1984):801–805.

8. World Bank, *The Challenge of Hunger;* World Bank, *Sub-Saharan*

Notes

Africa: From Crisis to Sustainable Growth, a Long-Term Perspective Study (Washington, D.C.: The World Bank, 1989).

9. World Bank, *The Challenge of Hunger*; P. Hendry, "Food and Population: Beyond Five Billion," in *Population Bulletin* (Washington, D.C.: Population Reference Bureau, 1988), Vol. 43; United Nations World Food Council, *The Global State of Hunger and Malnutrition: 1988 Report* (New York: United Nations World Food Council, 1988); J. von Braun and L. Paulino, "Food in Sub-Saharan Africa: Trends and Policy Challenges for the 1990s," *Food Policy*, 15(1990):505–517.

10. D. Ghai and L. Smith, *Agricultural Prices, Policy, and Equity in Sub-Saharan Africa* (Boulder, CO: Lynne Reinner Publishers, 1987); A. Hansen and D. McMillan, *Food in Sub-Saharan Africa* (Boulder, CO: Lynne Reinner Publishers, 1986); Huss-Ashmore and Katz, *African Food Systems;* P. Pinstrup-Andersen, *Government Policy, Food Security and Nutrition in Sub-Saharan Africa* (Ithaca, NY: Food and Nutrition Policy Program, Cornell University, 1989); World Resources Institute, *World Resources Report 1991* (New York: Oxford University Press, 1991).

11. M. K. Tolba, "Our Biological Heritage Under Siege," *BioScience*, 39(1989):725–728.

12. Mabbutt, "Impacts of Carbon Dioxide."

13. Food and Agriculture Organization, *Land, Food and People* (Rome: Food and Agriculture Organization, 1984); P. Harrison, *The Greening of Africa* (London: Palladin Books, 1987).

14. Food and Agriculture Organization, *Land, Food and People.*

15. McNamara, "Population and Africa's Development Crisis."

16. World Bank, *World Development Report 1991.*

17. Harrison, *The Greening of Africa;* J. W. Mellor, C. L. Delgado, and M. J. Blackie, eds., *Accelerating Food Production in Sub-Saharan Africa* (Baltimore: Johns Hopkins University Press, 1987); J. W. Mellor and S. Gavian, "Famine: Causes, Prevention and Relief," *Science*, 235(1987):539–545; J. Moris, "Irrigation as a Priority Solution in African Development," *Development Policy Review*, 5(1987):99–123.

18. M. Duffield, *War and Famine in Africa* (Oxford: Oxfam, 1990); N. Myers, "The Environmental Dimension to Security Issues," *The Environmentalist*, 6(4, 1986):251–257.

19. N. Birdsall and F. T. Sai, "Family Planning Services in Sub-Saharan Africa," *Finance and Development*, March (1988):28–31.

20. Diallo, *Population Policies;* McNamara, "Population and Africa's Development Crisis."

21. R. Gulhati, *The Political Economy of Reform in Sub-Saharan Africa* (Washington, D.C.: Economic Development Institute, The World Bank, 1988); J. P. Lassoie and S. Kyle, *Policy Reform and Natural Resources Man-*

agement in Sub-Saharan Africa (Ithaca, NY: Departments of Natural Resources and Agricultural Economics, Cornell University, 1989); World Bank, *The Challenge of Hunger.*

22. J. P. Grant, *The State of the World's Children 1989* (New York: UNICEF, 1989).

23. M. H. Glantz and R. W. Katz, "African Drought and Its Impacts: Revived Interest in a Current Phenomenon," *Desertification Control Bulletin*, 14(1987):22–30; see also United Nations Environment Programme, *Desertification Control in Africa: Actions and Directions of Institutions* (Nairobi: United Nations Environment Programme, 1985).

24. H. E. Dregne and C. J. Tucker, "Desert Encroachment," *Desertification Control Bulletin*, 16(1988):16–19.

CHAPTER 6: The Philippines

1. T. Abate, "Into the Northern Philippine Rainforests," *BioScience*, 42(1992):246–251; G. Cavanagh and R. Broad, *Plundering Paradise: People, Power and the Struggle Over the Environment in the Philippines* (Berkeley, CA: University of California Press, 1992); C. Fay, *Counter Insurgency and Tribal Peoples in the Philippines* (Washington, D.C.: Survival International U.S.A., 1987); J. B. Goodno, *The Philippines: Land of Broken Promises* (London: Zed Books, 1991); L. Nash, *Strategic Insecurity: Environment, Development, and Geopolitics in the Philippines* (Berkeley, CA: Pacific Institute for Studies in Development, Environment, and Security, 1991).

2. F. S. Factoran, "Population, Resources and the Future of the Philippines," *Populi*, 17(4, 1990):20–29; N. Myers, "Environmental Degradation and Some Economic Consequences in the Philippines," *Environmental Conservation*, 15(3, 1988):205–214; M. de los Angeles and M. Lasmarias, *A Review of Philippine Natural Resource and Environmental Management* (Manila, Philippines: Philippine Institute for Development Studies, 1990); G. Porter and J. Ganapin, *Resources, Population and the Philippines' Future* (Washington, D.C.: World Resources Institute, 1988); World Bank, *Population Pressure: The Environment and Agricultural Intensification in the Philippines.* (Washington, D.C.: The World Bank, 1990).

3. W. P. David, *Soil Erosion and Soil Conservation Planning—Issues and Implications* (Los Banos, Philippines: College of Engineering and Agro-Industrial Technology, University of the Philippines, 1987).

4. D. J. Ganapin, "Forest Resources and Timber Trade in the Philippines," in *Proceedings of the Conference on Forest Resources Crisis in the Third World* (Kuala Lumpur: Sahabat Alam Malaysia, 1987), pp. 54–70; D. M. Kummer, *Deforestation in the Post-War Philippines* (Boston: Department of Geography, Boston University, 1989); A. V. Revilla, *Policy and Program*

Agenda for the Forest Resources Management Sub-Sector (Los Banos, Philippines: College of Forestry, University of the Philippines, 1987).

5. Revilla, *Policy and Program Agenda.*

6. M. C. J. Cruz, *Population Pressure in Tropical Developing Countries* (Washington, D.C.: Population Reference Bureau, 1991); W. D. Cruz and M. C. J. Cruz, "Population Pressure and Deforestation in the Philippines," *ASEAN Economic Bulletin,* 7(1991):200–212.

7. Factoran, "Population, Resources and the Future."

8. G. M. Bautista, "The Forestry Crisis in the Philippines: Nature, Causes and Issues," *The Developing Economies,* 28(1990):1; B. T. Leong and C. B. Serna, *Status of Watersheds in the Philippines* (Quezon City, Philippines: National Irrigation Administration, 1987).

9. Cruz and Cruz, *Population Pressure.*

10. World Bank, *Population Pressure.*

11. Conservation of Fisheries and Aquatic Resources Task Force, *Issues, Problems, and Recommendations in Management and Conservation of Fisheries and Aquatic Resources* (Manila, Philippines: Department of Environment and Natural Resources, 1987); G. T. Silvestre, *Philippine Marine Capture Fisheries—Exploitation, Potential and Options for Sustainable Development* (Visayas, Diliman, Philippines: College of Fisheries, University of the Philippines, 1987).

12. G. T. Silvestre and S. Ganaden, *Status of Philippine Demersal Stocks: An Overview* (Visayas, Diliman, Philippines: College of Fisheries, University of the Philippines, 1987).

13. UNICEF, *State of the World's Children 1991* (New York: UNICEF, 1991).

14. R. L. Prosterman et al., eds., *Agrarian Reform and Grassroots Development: Ten Case Studies* (Boulder, CO: Lynne Reinner Publishers, 1990).

15. Cavanagh and Broad, *Plundering Paradise;* Goodno, *The Philippines.*

16. Fay, *Counter Insurgency;* Goodno, *The Philippines.* T. F. Homer-Dixon, J. H. Boutwell, and G. W. Rathjens, "Environmental Change and Violent Conflict," *Scientific American,* 268(2, 1993):38–45.

17. Abate, "Northern Philippine Rainforests"; Goodno, *The Philippines;* G. Hawes, "Theories of Peasant Revolution: A Critique and Contribution from the Philippines," *World Politics,* 42(1990):281–290; G. R. Jones, *Red Revolution: Inside the Philippine Guerilla Movement* (Boulder, CO: Westview Press, 1989).

18. A. Sfeir-Younis, *Soil Conservation in Developing Countries: A Background Report* (Washington, D.C.: The World Bank, 1986).

19. J. Spears and E. S. Ayensu, "Resources, Development, and the New Century: Forestry," in R. Repelto, ed., *The Global Possible* (New Haven: Yale University Press, 1985), pp. 299–336.

---

20. World Bank and World Resources Institute, *Tropical Forests: A Call for Action* (Washington, D.C.: World Bank and World Resources Institute, 1985).

21. N. Myers, *Deforestation Rates in Tropical Forests and Their Climatic Implications* (London: Friends of the Earth, 1989); N. Myers, "Tropical Forests: The Policy Challenge," *The Environmentalist*, 12(1, 1992):15–27.

22. R. E. Bilsborrow and M. E. Geores, *Population, Environment and Sustainable Agricultural Developments* (Chapel Hill, NC: Carolina Population Center, University of North Carolina, 1990); J. O. Browder, ed., *Fragile Lands of Latin America: Strategies for Sustainable Development* (Boulder, CO: Westview Press, 1989); M. C. J. Cruz and I. Zosa-Feranil, *Policy Implications of Population Pressure in the Philippine Uplands* (Los Banos, Philippines: Department of Environmental Studies, University of the Philippines, 1987); J. F. Hicks, H. E. Daly, S. H. Davis et al., *Ecuador's Amazon Region: Development Issues and Options* (Washington, D.C.: World Banks Discussion Papers 75, The World Bank, 1990); O. J. Lynch, *Whither the People? Demographic, Tenurial and Agricultural Aspects of the Tropical Forestry Action Plan* (Washington, D.C.: World Resources Institute, 1990); W. Manshard and W. B. Morgan, eds., *Agricultural Expansion and Pioneer Settlements in the Humid Tropics* (Tokyo: United Nations University, 1988); N. Myers, "The World's Forests and Human Populations: The Environmental Interconnections," in K. Davis and M. S. Bernstam, eds., *Resources, Environment and Population: Present Knowledge, Future Options* (New York: Oxford University Press, 1991), pp. 237–251; W. J. Peters and L. F. Neuenschwander, *Slash and Burn Farming in Third World Forests* (Moscow, ID: University of Idaho Press, 1988); R. L. Prosterman et al., eds., *Agrarian Reform and Grassroots Development: Ten Case Studies* (Boulder, CO: Lynne Reinner Publishers, 1990); D. Schuman and W. L. Partridge, *Human Ecology of Tropical Land Settlement in Latin America* (Boulder, CO: Westview Press, 1989); D. Southgate, "The Causes of Land Degradation on 'Spontaneously' Expanding Agricultural Frontiers in the Third World," *Land Economics*, 66(1990):93–101; W. C. Thiesenhusen, ed., *Searching for Agrarian Reform in Latin America* (Boston: Unwin Hyman, 1989).

23. Thiesenhusen, *Agrarian Reform*; N. MacDonald, *A Mask Called Progress* (Oxford: Oxfam, 1991).

Chapter 7: The Indian Subcontinent

1. N. Myers, "Environmental Security: The Case of South Asia," *International Environmental Affairs*, 1(2, 1989):138–154.

2. World Bank, *World Development Report 1991* (Washington, D.C.: The World Bank, 1991).

3. N. Myers, "Environmental Repercussions of Deforestation in the Himalayas," *Journal of World Forest Resource Management,* 2(1986):63–72.

4. A. N. Pandey, P. C. Pathak, and J. S. Singh, "Water, Sediment and Nutrient Movement in Forested and Non-Forested Catchments in Kumaun Himalaya," *Forest Ecology and Management,* 7(1984):19–29.

5. J. S. Singh, A. N. Pandey, and P. C. Pathak, "A Hypothesis to Account for the Major Pathway of Soil Loss from the Himalayas," *Environmental Conservation,* 10(1983):343–345.

6. S. Ahmad, M. Yasin, and G. R. Sandhu, *Efficient Irrigation Management Supplements: Land Drainage* (Islamabad: Pakistan Agricultural Research Council, 1987).

7. S. M. H. Bokhari, "Case Study on Waterlogging and Salinity Problems in Pakistan," *Water Supply and Management,* 4(1980):171–192.

8. Government of Pakistan, *Seventh Five-Year Plan 1988–93* (Islamabad: Ministry of Economic Planning, Government of Pakistan, 1988).

9. B. A. Malik, *Pakistan's Limited Water Resources and Growing Demand* (Islamabad: International Commission on Irrigation and Drainage, 1986).

10. S. M. Khan, "Management of River and Reservoir Sedimentation in Pakistan," *Water Resources Journal,* 149(1986):40–43.

11. C. P. R. Nottidge, G. Schreiber, and A. Q. Sheikh, *Pakistan: Environmental Rehabilitation, Protection and Management* (Islamabad: The World Bank, 1985).

12. R. Schwass, *National Conservation Strategy for Pakistan: Preliminary Appraisal* (Toronto: Faculty of Environmental Studies, York University, 1986).

13. M. I. Sheikh, personal communication, letter of 17 May 1987, Office of the Director-General of the Pakistan Forestry Institute, Peshawar, Pakistan.

14. M. Stetson, "People Who Live in Green Houses," *World Watch,* 4(5, 1991):22–29.

15. Pakistan Forestry Institute, *Forestry, Watershed, Range and Wildlife Management in Pakistan* (Peshawar: Pakistan Forestry Institute, 1983).

16. Government of Pakistan, *Five-Year Plan 1988–93.*

17. Pakistan Forestry Institute, *Forestry, Watershed, Range.*

18. K. Jalees, "Loss of Productive Soil in India," *International Journal of Environmental Studies,* 24(1985):245–250.

19. A. K. Biswas, "Environmental Concerns in Pakistan, with Special Reference to Water and Forests," *Environmental Conservation,* 14(1987):319–328.

20. Schwass, *Conservation Strategy for Pakistan.*

21. Government of Pakistan, *Environmental Profile of Pakistan* (Islamabad: Environment and Urban Affairs Division, Government of Pakistan, 1986).

22. S. K. Sinha, "The 1982–83 Drought in India: Magnitude and Impact," in M. Glantz, R. Katz, and M. Krenz, eds., *The Societal Impacts Associated with the 1982–83 Worldwide Anomalies* (Boston: Stan Grossfield/*Boston Globe*, 1987), pp. 37–42.

23. A. Agarwal and S. Narain, *Towards Green Villages* (Kilash Colony, New Delhi: Center for Science and Environment, 1990).

24. Ibid.

25. High Level Committee on Floods, Government of India, *Report of the Emergent Crisis* (New Delhi: High Level Committee on Floods, Government of India, 1983).

26. D. Seckler and R. K. Sampath, *Production and Poverty in Indian Agriculture* (Fort Collins, CO: International School for Agricultural and Resource Development, Colorado State University, 1985).

27. Ibid.

28. J. Bandyopadhyay, "Riskful Confusion of Drought and Man-Induced Water Scarcity," *Ambio*, 18(1989):284–292.

29. G. R. Choudhury and T. A. Khan, "Developing the Ganges Basin," in M. Zaman et al., eds., *River Basin Development* (Dublin: Tycooly Press, 1983).

30. B. B. Vohra, *Management of Natural Resources: Urgent Need for Fresh Thinking* (New Delhi: Office of the Chairman, Government Advisory Board on Energy, 1986).

31. D. R. Bhumba and A. Khare, *Estimate of Wastelands in India* (New Delhi: Society for Promotion of Wastelands Development, 1987).

32. M. L. Dewan and S. Sharma, *People's Participation as a Key to Himalayan Ecosystem Development* (New Delhi: Center for Policy Research, 1985).

33. A. Agarwar et al., *The Wrath of Nature: The Impact of Environmental Destruction on Floods and Droughts* (New Delhi: Center for Science and Environment, 1987).

34. Indian National Institute of Hydrology, *Forest Influences on Hydrological Parameters: Status Report* (Roorkee, India: Indian National Institute of Hydrology, 1986).

35. High Level Committee on Floods, Government of India, *Report of the Emergent Crisis.*

36. A. Sfeir-Younis, *Soil Conservation in Developing Countries—A Background Report* (Washington, D.C.: The World Bank, 1986).

37. Jalees, "Loss of Productive Soil"; V. V. D. Narayana and R. Babu, "Estimate of Soil Erosion in India," *Journal of Irrigation and Drainage Engineering*, 109(4, 1983):419–434.

38. World Resources Institute, *World Resources 1992–93* (New York: Oxford University Press, 1992).

39. Vohra, *Natural Resources.*

40. D. C. Das, Y. P. Bali, and R. N. Kaul, "Soil Conservation in Multiple Purpose River Valley Catchments," *Indian Journal of Soil Conservation,* 9(1981):6–20; Ministry of Agriculture, Government of India, *The State of India's Agriculture 1982* (New Delhi: Ministry of Agriculture, 1983).

41. Jalees, "Loss of Productive Soil."

42. Food and Agriculture Organization, *Agriculture: Toward 2000* (Rome: Food and Agriculture Organization, 1981); F. H. Sanderson, "World Food Prospects to the Year 2000," *Food Policy,* 9(1984):363–373.

43. M. S. Swaminathan and S. K. Sinha, eds., *Global Aspects of Food Production* (Dublin: Tycooly International Press, 1986); World Bank, *World Development Report 1992* (New York: Oxford University Press, 1992).

44. A. Agarwal and S. Narain, *The State of India's Environment 1984–85* (New Delhi: Center for Science and Environment, 1985); Sinha, "The 1982–83 Drought."

45. Jalees, "Loss of Productive Soil."

46. Agarwal and Narain, *Towards Green Villages.*

47. Seckler and Sampath, *Production and Poverty;* see also Vohra, *Natural Resources.*

48. Swaminathan and Sinha, *Global Aspects.*

49. F. U. Mahtab and Z. Karim, "Population and Agricultural Land Use: Towards a Sustainable Food Production System in Bangladesh," *Ambio,* 21(1, 1992):50–55.

50. World Bank, *World Development Report 1990* (Washington, D.C.: The World Bank, 1990).

51. E. G. Jansen, *Rural Bangladesh: Competition for Scarce Resources* (Bergen, Norway: Christian Michelson Institute, 1983).

52. R. S. McNamara, "The Population Problem: Time Bomb or Myth?" *Foreign Affairs,* 62(1984):1107–1131.

53. S. R. Millman, R. S. Chen, J. Emlen et al., *The Hunger Report: Update 1991* (Providence, RI: World Hunger Program, Brown University, 1991); World Bank, *World Development Report 1992.*

54. I. Singh, *Small Farmers and the Landless in South Asia* (Washington, D.C.: The World Bank, 1981); R. Sinha, *Landlessness: A Growing Problem* (Rome: Food and Agriculture Organization, 1984); T. N. Srinivasan and P. K. Bardhan, eds., *Rural Poverty in South Asia* (New York: Columbia University Press, 1988).

55. R. L. Sivard, *World Military and Social Expenditures 1991* (Washington, D.C.: World Priorities Inc., 1991).

Chapter 8: El Salvador

1. President Ronald Reagan, speech to joint session of Congress, 27 April 1983, Washington, D.C.

2. S. Hilty, *Environmental Profile of El Salvador* (Tucson: Arid Lands Information Center, University of Arizona, 1982).

3. U.S. Agency for International Development, *Environmental Profile of El Salvador* (Washington, D.C.: Bureau of Science and Technology, U.S. Agency for International Development, 1982).

4. H. Kissinger et al., *Report of the President's National Bipartisan Commission on Central America* (The Kissinger Commission), January 1984 (Washington, D.C.: U.S. Government Printing Office, 1984).

5. H. J. Leonard, *Natural Resources and Economic Development in Central America* (New Brunswick, NJ: Transaction Books, 1987); H. J. Wiarda, ed., *Rift and Revolution: The Central American Imbroglio,* (Washington, D.C.: American Enterprise Institute, 1986); S. Annis, ed., *Poverty, Natural Resources, and Public Policies in Central America* (Washington, D.C.: ODC, 1992).

6. W. Hall and D. Faber, *El Salvador: Ecology of Conflict* (San Francisco: Environmental Project on Central America, 1989).

7. World Bank, *World Development Report 1991* (Washington, D.C.: The World Bank, 1991).

8. J. Karliner, "Central America's Other War," *World Policy Journal,* 6(1989):787–810.

9. M. Diskin, ed., *Trouble in Our Backyard: Central America and the United States in the Eighties* (New York: Pantheon Books, 1983); D. Paarlberg, P. M. Cody, and R. J. Ivey, *Agrarian Reform in El Salvador* (Washington, D.C.: Checci and Company, 1986); R. Sinha, *Landlessness: A Growing Problem* (Rome: Food and Agriculture Organization, 1984).

10. Leonard, *Natural Resources.*

11. L. Garcia, *Analysis of Watershed Management: El Salvador, Guatemala, Honduras* (Washington, D.C.: U.S. Agency for International Development, 1982).

12. H. J. Leonard, "Managing Central America's Renewable Resources: The Path to Sustainable Economic Development," *International Environmental Affairs,* 1(1, 1990):38–56; S. G. Lustgarden, *El Salvador: The Political Economy of Environmental Destruction* (San Francisco: Environmental Project on Central America, 1984); G. Wirick, "Environment and Security: The Case of Central America," *Peace and Security,* 4(1989):2–3.

13. L. North, *Bitter Grounds: Roots of Revolt in El Salvador* (Toronto: Between the Lines Publishers, 1985).

14. T. P. Anderson, *The War of the Dispossessed: Honduras and El Salvador 1969* (Lincoln, NE: University of Nebraska Press, 1981); W. H. Durham, *Scarcity and Survival in Central America: Ecological Origins of the Soccer War* (Stanford, CA: Stanford University Press, 1979).

15. S. Montes, "Migration to the United States as an Index of the Intensifying Social and Political Crises in El Salvador," *Journal of Refugee Studies,*

Notes

1(1988):107–126; S. M. Mozo and J. J. V. Vasquez, *Salvadoran Migration to the United States: An Exploratory Study* (Washington, D.C.: Center for Immigration Policy and Refugee Assistance, Georgetown University, 1988).

16. Hall and Faber, *El Salvador*.

17. L. R. Simon and J. C. Stephens, *El Salvador Land Reform 1980–81: Impact Audit* (Boston: Oxfam America, 1982); U.S. AID Inspector General (Latin America), *Agrarian Reform in El Salvador: Report on Its Status, 1984* (Washington, D.C.: U.S. Agency for International Development, 1988); U.S. Agency for International Development, *Environment and Natural Resources: A Strategy for Central America* (Washington, D.C.: U.S. Agency for International Development, 1989).

18. Food and Agriculture Organization, *Potential Population Supporting Capacities of Lands in the Developing World* (Rome: Food and Agriculture Organization, 1984); P. Harrison, *Land, Food and People* (Rome: Food and Agriculture Organization, 1984).

19. World Bank, *World Development Report 1991* (Washington, D.C.: The World Bank, 1991).

20. Arms Control and Foreign Policy Caucus of the U.S. Congress, *Bankrolling Failure: U.S. Policy in El Salvador and the Urgent Need for Reform* (Washington, D.C.: U.S. Congress, 1987); M. Klare and P. Kornbluh, *American Military Policy in Small Wars: The Case of El Salvador* (Cambridge, MA: John F. Kennedy School of Government, Harvard University, 1988).

21. J. C. Stephens, *El Salvador: Environment and War* (Philadelphia: American Friends Service Committee, 1989).

22. K. Coleman and G. Herring, eds., *The Central American Crisis: Sources of Conflict and the Failure of U.S. Policy* (Wilmington, DE: Scholarly Resources Inc., 1985); R. S. Leiken, ed., *Central America: Anatomy of Conflict* (New York: Pergamon Press, 1984); S. C. Stonich, "The Dynamics of Social Processes and Environmental Destruction: A Central American Case Study," *Population and Development Review*, 15(1989):269–296.

23. T. Barry and D. Preusch, *The Central America Fact Book* (New York: Grove Press, 1986); Leonard, *Natural Resources*.

24. W. J. Weinberg, *War on the Land: Ecology and Politics in Central America* (London: Zed Books, 1991).

25. Leonard, *Natural Resources*.

26. World Bank, *World Development Report 1990* (Washington, D.C.: The World Bank, 1990).

27. Leonard, *Natural Resources*.

28. R. Hough et al., *Land and Labor in Guatemala: An Assessment* (Washington, D.C.: U.S. Agency for International Development, 1988).

29. J. L. Posner and M. F. McPherson, "Agriculture on the Steep Slopes of Tropical America: The Current Situation and Prospects," *World Development*, May (1982):341–354.

30. T. Barry, *Roots of Rebellion, Land and Hunger in Central America* (Boston: Southend Press, 1987).

31. R. L. Sivard, *World Military and Social Expenditures 1991* (Washington, D.C.: World Priorities Inc., 1991); Wirick, "Environment and Security."

32. P. W. Fagen and S. Aguayo, *Fleeing the Maelstrom: Central American Refugees* (Washington, D.C.: School of Advanced International Studies, Johns Hopkins University, 1986); R. E. Feinberg and C. R. Carlisle, *Immigration to the United States from Central America: Some Thoughts on Its Causes and Cures* (Washington, D.C.: Commission for the Study of International Migration and Cooperative Economic Development, U.S. Congress, 1989); D. Gallagher and J. M. Diller, *At the Crossroads between Uprooted People and Development in Central America* (Washington, D.C.: Commission for the Study of International Migration and Cooperative Economic Development, U.S. Congress, 1990).

33. A. Berryman and P. Berryman, *In the Shadow of Liberty: Central American Refugees in the United States* (Philadelphia: American Friends Service Committee, 1988); G. J. Borjas, *Friends or Strangers: The Impact of Immigration on the U.S. Economy* (New York: Basic Books, 1991); M. Glazer, *The New Immigration: A Challenge to American Society* (San Diego: San Diego State University Press, 1988).

34. L. A. Lewis and W. J. Coffey, "The Continuing Deforestation of Haiti," *Ambio*, 14(1985):158–160.

35. World Bank, *Haiti Agricultural Sector Study* (Washington, D.C.: The World Bank, 1985).

36. A. Waldman and C. Foster, eds., *Haiti: Today and Tomorrow* (Lanham, MA: University of America Press, 1984).

37. International Development Association, *Natural Resource Management for Durable Development in Haiti* (Washington, D.C.: International Development Association, 1990).

38. Leonard, *Natural Resources.*

39. R. Tata, *Haiti: Land of Poverty* (New York: University Press of America, 1982).

40. A. V. Catanese, *Haiti's Refugees: Political, Economic, Environmental* (Washington, D.C.: Universities Field Staff International and the Natural Heritage Institute, 1991).

41. Caribbean Migration Program, *Haitian Migration and the Haitian Economy* (Gainesville, FL: Center for Latin American Studies, University of Florida, 1984); J. DeWind and D. Kinley, *Aiding Migration: The Impact of International Development Assistance on Haiti* (Boulder, CO: Westview Press, 1988); A. Stepick, "Haitian Boat People: A Study in Conflicting Forces Shaping U.S. Immigration Policy," *Law and Contemporary Problems,* 45(1982):163–196.

42. D. Barker, *Environmental Migrants* (New York: United Nations Development Program, 1989); Catanese, *Haiti's Refugees.*

43. K. Danaher, P. Berryman, and M. Benjamin, *Help or Hindrance? U.S. Economic Aid in Central America* (San Francisco: Institute for Food and Development Policy, 1987).

44. T. J. Espenshade, "Growing Imbalances between Neighbor Supply and Labor Demand in the Caribbean Basin," in F. D. Bean, J. Schmandt, and S. Weintraub, eds., *Mexican and Central American Population and U.S. Immigration Policy* (Austin, TX: Center for Mexican-American Studies, University of Texas, 1989), pp. 113–160.

45. Feinberg and Carlisle, *Immigration to the United States from Central America.*

46. C. A. Quesada, *Population, Development, Resources and Environment Linkages: A Costa Rican Perspective* (San Jose, Costa Rica: Secretariat of the Costa Rican Conservation Strategy for Sustainable Development, 1989).

47. J. P. Augelli, "Costa Rica: Transition to Land Hunger and Potential Instability," *1984 Yearbook of Conference of Latin Americanist Geographers,* 10(1984):48–61.

48. J. Terborgh, *Where Have All the Birds Gone?* (Princeton, NJ: Princeton University Press, 1989).

CHAPTER 9: Mexico

1. W. D. Rogers, "Approaching Mexico," *Foreign Policy,* 72(1988):196–209.

2. I. Restrepo, *Naturaleza Muerto: Los Plaguicidas en Mexico* (Mexico City: Ediciones Oceano, 1988); A. Sherbinin, "Survey Report: Mexico," *Population Today,* 18(1990):5.

3. R. G. Cummings, *Improving Water Management in Mexico's Irrigated Agricultural Sector* (Washington, D.C.: World Resources Institute, 1989).

4. J. A. Mabbutt, "A New Global Assessment of the Status and Trends of Desertification," *Environmental Conservation,* 11(1984):103–113.

5. D. M. Liverman, "Environment and Security in Mexico," in S. Aguayo and M. Bagley, eds., *Issues in Mexican National Security* (University Park, PA: Pennsylvania State University Press, 1993).

6. N. Myers, *Future Operational Monitoring of Tropical Forests: An Alert Strategy* (Ispra, Italy: Ispra Joint Research Center, Commission of the European Community, 1992).

7. M. Redclift, "Mexico's Green Movement," *The Ecologist,* 17(1, 1987):44–46; V. M. Toledo, *Ecologia y Autosuficiencia Alimentaria* (Mexico City: Siglo Veintiuno Editores, 1985); Restrepo, *Naturaleza Muerto.*

8. M. L. Carlos, *State Policies, State Penetration and Ecology: A Compara-*

tive *Analysis of Uneven Development and Underdevelopment in Mexico's Micro-Agrarian Regions* (La Jolla/San Diego: Center for U.S.–Mexican Studies, University of California at San Diego, 1981); W. E. Doolittle, "Agricultural Expansion in Marginal Areas of Mexico," *Geographical Review*, 73(1983):301–313.

9. S. Sanderson, *The Transformation of Mexican Agriculture: International Structure and the Politics of Rural Change* (Princeton, NJ: Princeton University Press, 1986); M. S. Grindle, *Official Interpretations of Rural Underdevelopment: Mexico in the 1970s* (La Jolla/San Diego: Program in U.S.–Mexican Studies, University of California at San Diego, 1981).

10. S. A. Quezada and B. M. Bagley, eds., *En Busca de la Seguridad Perdida: Aproximaciones a la Seguridad Nacional Mexicana* (Mexico City: Siglo Veintiuno Editones, 1990).

11. G. D. Thompson and P. L. Martin, *The Potential Effects of Labor-Intensive Agriculture in Mexico on United States–Mexico Migration* (Washington, D.C.: Commission for the Study of International Migration and Cooperative Economic Development, U.S. Congress, 1989).

12. World Bank, *World Development Report 1992* (Washington, D.C.: The World Bank, 1991).

13. World Bank, *World Development Report 1992* (Washington, D.C.: The World Bank, 1992).

14. D. M. Liverman, *Environment and Security in Mexico* (University Park, PA: Department of Geography, Pennsylvania State University, 1991); Toledo, *Ecologia;* S. Whiteford and L. Montgomery, "The Political Economy of Rural Transformation: The Mexican Case," in B. R. DeWalt and P. J. Pelto, eds., *Micro and Macro Levels of Analysis in Anthropology: Issues in Theory and Research* (Boulder, CO: Westview Press, 1984), 146–164.

15. W. A. Cornelius, *The Cactus Curtain* (Berkeley, CA: University of California Press, 1993); W. A. Cornelius and J. A. Bustamante, eds., *Mexican Migration to the United States* (La Jolla/San Diego: Center for U.S.–Mexican Studies, University of California at San Diego, 1989); S. Diaz-Briquets and S. Weintraub, eds., *Determinants of Migration from Mexico, Central America and the Caribbean* (Boulder, CO: Westview Press, 1991); A. Schumacher, "Agricultural Development and Rural Employment: A Mexican Dilemma," *Working Papers in U.S.–Mexican Studies*, 21 (La Jolla/San Diego: Program in United States–Mexican Studies, University of California at San Diego, 1981).

16. Thompson and Martin, *Labor-Intensive Agriculture;* Liverman, *Environment and Security;* G. Vernez and D. Ronfeldt, "The Current Situation in Mexican Immigration," *Science*, 251(1991):1189–1193.

17. World Resources Institute, *World Resources Report 1990* (New York: Oxford University Press, 1990).

18. F. Alba and J. E. Potter, "Population and Development in Mexico

Since 1940: An Interpretation," *Population and Development Review,* 12(1986):47–75; Sherbinin, "Survey Report: Mexico."

19. Diaz-Briquets and Weintraub, *Migration from Mexico.*

20. Liverman, "Environment and Security."

21. L. Bouvier and D. Simcox, *Many Hands, Few Jobs: Population, Unemployment and Emigration in Mexico and the Caribbean* (Washington, D.C.: Center for Immigration Studies, 1986); J. G. Castenada, "Mexico at the Brink," *Foreign Affairs,* 64(1985):287–303.

22. R. Sinha, *Landless: A Growing Problem* (Rome: Food and Agriculture Organization, 1984).

23. W. D. Rogers, "Approaching Mexico," *Foreign Policy,* 72(1988):196–209.

24. R. S. Chen, W. H. Bender, R. W. Kates et al., *The Hunger Report: 1990* (Providence, RI: World Hunger Program, Brown University, 1990).

25. Castenada, "Mexico at the Brink"; M. Gendell, "Population Growth and Labor Absorption in Latin America, 1970–2000," in J. Saunders, ed., *Population Growth in Latin America and U.S. National Security* (Boston: Allen and Unwin, 1986), pp. 49–78.

26. Castenada, "Mexico at the Brink"; S. Trejo-Reyes, "Mexico's Long Travail," *Development Forum,* 15(1987):3, 10.

27. T. Espenshade, *The Cost of Job Creation in the Caribbean* (Washington, D.C.: The Urban Institute, 1987); T. J. Espenshade, "Growing Imbalances between Neighbor Supply and Labor Demand in the Caribbean Basin," in F. D. Bean, J. Schmandt, and S. Weintraub, eds., *Mexican and Central American Population and U.S. Immigration Policy* (Austin, TX: Center for Mexican American Studies, University of Texas, 1989), pp. 113–160.

28. F. D. Bean, J. Schmandt, and S. Weintraub, eds., *Mexican and Central American Population and U.S. Immigration Policy* (Austin, TX: University of Texas Press, 1989); R. A. Pastor and J. Castenada, *Limits to Friendship: The United States and Mexico* (New York: Knopf, 1988).

29. Bilateral Commission on the Future of United States–Mexican Relations, *The Challenge of Interdependence: Mexico and the United States* (Lanham, MD: University Press of America, 1988); L. Werner, "Population and Migration: A Case from Mexico," *Populi,* 18(1991):46–51.

30. G. J. Borjas, *Friends or Strangers: The Impact of Immigration on the U.S. Economy* (New York: Basic Books, 1991); B. Edmonston and J. S. Passell, eds., *Immigration and Ethnicity: The Adjustment of America's Newest Immigrants* (Washington, D.C.: The Urban Institute, 1992); J. Saunders, ed., *Population Growth in Latin America and U.S. National Security* (Boston: Allen and Unwin, 1986).

31. L. F. Bouvier, *Peaceful Invasions: Immigration and Changing America* (Washington, D.C.: Center for Immigration Studies, 1991).

32. Bouvier, *Peaceful Invasions;* S. H. Preston, *Demographic Change in the*

United States 1970–2050 (Philadelphia: The Wharton School, University of Pennsylvania, 1991).

33. Bouvier, *Peaceful Invasions;* Edmonston and Passell, *Immigration and Ethnicity;* Preston, *Demographic Change.*

34. Pastor and Castenada, *Limits to Friendship;* Rogers, "Approaching Mexico."

35. Bouvier and Simcox, *Many Hands.*

36. Liverman, *Environment and Security.*

CHAPTER 10: Population

1. N. Choucri, "Demographics and Conflict," *Bulletin of the Atomic Scientists,* 42(1986):24–25; N. Choucri and R. C. North, "Lateral Pressure in International Relations: Concept and Theory," in M. I. Midlarsky, ed., *Handbook of War Studies* (Winchester, MA: Unwin Hyman Inc., 1989); P. R. Ehrlich and A. H. Ehrlich, *The Population Explosion* (New York: Simon and Schuster, 1990); J. E. Harf and B. T. Trout, *The Politics of Global Resources: Energy, Environment, Population and Food* (Durham, NC: Duke University Press, 1986); R. S. McNamara, "The Population Problem: Time Bomb or Myth?" *Foreign Affairs,* 62(1984):1107–1131; R. S. McNamara, *A Global Population Policy to Advance Human Development in the 21st Century* (New York: United Nations, 1991); N. Myers, "Population, Environment and Conflict," *Environmental Conservation,* 14(1, 1987):15–22; J. Saunders, ed., *Population Growth in Latin America and U.S. National Security* (Boston: Allen and Unwin, 1986).

2. Ehrlich and Ehrlich, *Population Explosion;* McNamara, "Population Problem"; McNamara, *Global Population Policy.*

3. United Nations Population Division, *Long-Range World Population Projections: Two Centuries of Population Growth* (New York: United Nations Population Division, 1991).

4. United Nations Population Division, *World Population Projections;* World Bank, *World Development Report: Environment and Development* (Washington, D.C.: The World Bank, 1992).

5. World Bank, *World Development Report 1991* (Washington, D.C.: The World Bank, 1991).

6. Ibid.

7. Ehrlich and Ehrlich, *Population Explosion;* P. Harrison, *The Third Revolution: Environment, Population and a Sustainable World* (London: I. B. Tauris, and New York: St. Martin's Press, 1992); N. Keyfitz, "The Growing Human Population," *Scientific American,* 261(1989):119–126; N. Keyfitz, *Reconciling Economic and Ecological Theory on Population* (Laxenburg, Austria: International Institute for Applied Systems Analysis, 1989); N. Keyfitz, *Seven Ways of Causing the Less Developed Countries' Population Prob-*

lem to Disappear—In Theory (Laxenburg, Austria: International Institute for Applied Systems Analysis, 1991); N. Myers, *Population, Resources and the Environment: The Critical Challenges* (New York: United Nations Population Fund, 1991).

8. L. H. Brown et al., *State of the World 1990* (New York: W. W. Norton, 1990).

9. Ibid.

10. P. R. Ehrlich, G. C. Daily, A. H. Ehrlich et al., *Global Change and Carrying Capacity: Implications for Life on Earth* (Stanford, CA: Stanford Institute for Population and Resource Studies, Stanford University, 1989); see also H. E. Daly and J. B. Cobb, Jr., *For the Common Good* (Boston: Beacon Press, 1989); N. Keyfitz, *Population Growth Can Prevent the Development that Would Slow Population Growth* (Laxemburg, Austria: Population Programme, International Institute for Applied Systems Analysis, 1990); D. Pimentel and M. Pimentel, "Land, Energy and Water: The Constraints Governing Ideal U.S. Population Size," in *The NPG Forum* (Teaneck, NJ: Negative Population Growth Inc., 1989).

11. R. S. Chen, W. H. Bender, R. W. Kates et al., *The Hunger Report: 1990* (Providence, RI: World Hunger Program, Brown University, 1990).

12. Brown et al., *State of the World 1990.*

13. M. Lipton, *The Poor and the Poorest: Some Interim Findings* (Washington, D.C.: The World Bank, 1985); see also N. Sadik, *The State of World Population 1992: A World in Balance* (New York: United Nations Population Fund, 1992).

14. R. W. Kates and V. Haarmann, *Poor People and Threatened Environments: Global Overviews, Country Comparisons, and Local Studies* (Providence, RI: World Hunger Program, Brown University, 1991); Keyfitz, *Population Growth;* P. D. Little and M. M. Horowitz, eds., *Lands at Risk in the Third World: Local-Level Perspectives* (Boulder, CO: Westview Press, 1987); S. Mink, *Poverty, Population, and the Environment* (Washington, D.C.: The World Bank, 1991).

15. Keyfitz, *Population Growth.*

16. R. S. McNamara, "Population and Africa's Development Crisis," *Populati,* 17(4, 1990):20–29.

17. Pimentel and Pimentel, "Land, Energy and Water."

18. Harrison, *The Third Revolution;* Ehrlich and Ehrlich, *Population Explosion;* Myers, *Critical Challenges.*

CHAPTER 11: Ozone-Layer Depletion and Global Warming

1. R. E. Benedick, *Ozone Diplomacy: New Directions in Safeguarding the Planet* (Cambridge, MA: Harvard University Press, 1991); I. Mintzer, W. R. Moomaw, and A. S. Miller, *Protecting the Ozone Shield: Strategies for Phas-*

ing Out CFCs during the 1990s (Washington, D.C.: World Resources Institute, 1990).

2. R. C. Worrest and M. M. Caldwell, eds., *Stratospheric Ozone Reduction, Solar Ultraviolet Radiation and Plant Life* (New York: Springer-Verlag, 1986).

3. R. C. Worrest and D.-P. Hader, "The Effects of Stratospheric Ozone Depletion in Marine Organisms," *Environmental Conservation,* 16(1989): 261–263.

4. F. S. Rowland, "Stratospheric Ozone Depletion by Chlorofluorocarbons," *Ambio,* 19(1990):281–292.

5. Benedick, *Ozone Diplomacy.*

6. G. D. Phillips, "CFCs in the Developing Nations: A Major Economic Development Opportunity," *Ambio,* 19(1990):316–320; A. Rosencranz and R. Milligan, "CFCs Abatement: Needs of Developing Countries," *Ambio,* 19(1990):312–316.

7. I. Isaksen, L. Roke, and M. Fergus, *Report of the United Nations Development Programme Mission to Investigate Ozone Layer Protection in China* (Beijing: National Environmental Protection Agency, 1990).

8. Benedick, *Ozone Diplomacy;* C. Bruhl and P. J. Crutzen, "Ozone and Climate Changes in the Light of the Montreal Protocol: A Model Study," *Ambio,* 19(1990):293–301.

9. United Nations Environment Programme, *Ozone-Layer Depletion: Environmental Effects Panel Report* (Nairobi: United Nations Environment Programme, 1989).

10. W. H. Brune et al., "The Potential for Ozone Depletion in the Arctic Polar Stratosphere," *Science,* 252(1991):1260–1266.

11. Benedick, *Ozone Diplomacy.*

12. J. T. Houghton, G. J. Jenkins, and J. J. Ephramus, eds., *Climate Change: The IPCC Scientific Assessment* (Cambridge, U.K.: Cambridge University Press, 1990); J. T. Houghton et al., *1992 IPCC Supplement* (Geneva: World Meteorological Organization, and Nairobi: United Nations Environment Programme, 1992); J. Leggett, ed., *Global Warming: The Greenpeace Report* (Oxford: Oxford University Press, 1990); M. Oppenheimer and R. H. Boyle, *Dead Heat: The Race against the Greenhouse Effect* (New York: Basic Books, 1990); S. H. Schneider, *Global Warming: Are We Entering the Greenhouse Century?* (San Francisco: Sierra Club Books, 1989).

13. A. Gore, *Earth in the Balance* (Boston: Houghton Mifflin, 1992).

14. Houghton et al., *Climate Change;* Houghton et al., *1992 IPCC Supplement.*

15. J. P. Holdren, "Energy in Transition," *Scientific American,* 263(3, 1990):109–115.

16. U.S. Office of Technology Assessment, *Changing by Degrees: Steps to Reduce Greenhouse Gases* (Washington, D.C.: Office of Technology Assessment, 1991).

17. Schneider, *Global Warming.*

18. P. H. Gleick, "The Implications of Global Climatic Changes for International Security," *Climatic Change,* 15(1989):309–326.

19. G. K. Heilig, *The Greenhouse Gas Methane: Sources and Sinks, the Impact of Population Growth, Possible Interventions* (Laxenburg, Austria: International Institute for Applied Systems Analysis, 1992).

20. G. C. Daily and P. R. Ehrlich, *An Exploratory Model of the Impact of Rapid Climate Change on the World Food Situation* (Stanford, CA: The Morrison Institute for Population and Resource Studies, Stanford University, 1990); Schneider, *Global Warming.*

21. D. Pimentel et al., *Offsetting Global Climate Changes on Food Production* (Ithaca, NY: College of Agriculture, Cornell University, 1990).

22. C. Rosenzweig and M. Parry, *Implications of Climate Change for International Agriculture: Global Food Trade and Vulnerable Regions* (Washington, D.C.: Environmental Protection Agency, 1992).

23. For instance, Schneider, *Global Warming;* J. B. Smith and D. A. Tirpak, *The Potential Effects of Global Climate Change on the United States* (New York: Hemisphere Publishing Corporation, 1990).

24. Oppenheimer and Boyle, *Dead Heat;* Schneider, *Global Warming.*

25. Rosenzweig and Parry, *Implications of Climate.*

26. Houghton et al., *Climate Change;* Houghton et al., *1992 IPCC Supplement;* Rosenzweig and Parry, *Implications of Climate.*

27. J. J. Mackenzie, R. C. Dower, and D. D. T. Chen, *The Going Rate: What It Really Costs to Drive* (Washington, D.C.: World Resources Institute, 1992); D. Yergin, *Gasoline and the American People* (Cambridge, MA: Cambridge Energy Research Associates, 1991).

28. A. B. Lovins, *Profitably Abating Global Warming* (Snowmass, CO: Rocky Mountain Institute, 1991); Yergin, *Gasoline.*

29. W. Zuckerman, *End of the Road: The World Car Crisis and How We Can Solve It* (Coast Mills, VT: Chelsea Green Publishing Company, 1991).

30. Lovins, *Abating Global Warming.*

31. Zuckerman, *End of the Road;* A. Krupnick, *The Environmental Costs of Energy Supply: A Framework for Estimation* (Washington, D.C.: Resources for the Future, 1990).

32. R. Dower and M. B. Timmerman, *The Right Climate for Carbon Taxes: Creating Economic Incentives to Protect the Atmosphere* (Washington, D.C.: World Resources Institute, 1992).

33. H. M. Hubbard, "The Real Cost of Energy," *Scientific American,* 264(4, 1991):18–23; Lovins, *Abating Global Warming;* U.S. Office of Technology Assessment, *Changing by Degrees.*

34. Hubbard, "Real Cost"; MacKenzie et al., *Going Rate.*

35. Lovins, *Abating Global Warming.*

36. Ibid.

37. Ibid.

38. Ibid.

39. McKinsey and Company, *Protecting the Global Atmosphere: Funding Mechanisms* (Amsterdam: McKinsey and Company, 1989).

40. U.S. Office of Technology Assessment, *Fueling Development: Energy Technologies for Developing Countries* (Washington, D.C.: Office of Technology Assessment, 1992); see also J. C. Topping, *Global Warming: Impact on Developing Countries* (Washington, D.C.: Overseas Development Council, 1990).

CHAPTER 12: Mass Extinction of Species

1. P. R. Ehrlich and A. H. Ehrlich, *The Population Explosion* (New York: Simon and Schuster, 1990); N. Myers, "Threatened Biotas: 'Hot Spots' in Tropical Forests," *The Environmentalist*, 8(1988):187–208; N. Myers, "The Biodiversity Challenge: Expanded Hot-Spots Analysis," *The Environmentalist*, 10(4, 1990):243–256; P. R. Raven, "The Politics of Preserving Biodiversity," *BioScience*, 40(10, 1990):769–774; E. O. Wilson, *The Diversity of Life* (Cambridge, MA: Belknap Press, 1992).

2. N. Myers, *A Wealth of Wild Species* (Boulder, CO: Westview Press, 1983); M. L. Oldfield, *The Value of Conserving Genetic Resources* (Sunderland, MA: Sinauer Associates, 1989).

3. Myers, "Threatened Biotas"; Myers, "Biodiversity."

4. L. R. Nault and W. R. Findley, "Primitive Relative Offers New Traits for Corn Improvement," *Ohio Report*, 66(1982):90–92.

5. A. C. Fisher, *Economic Analysis and the Extinction of Species* (Berkeley, CA: Department of Agriculture and Resource Economics, University of California, 1982); Myers, *Wild Species*.

CHAPTER 13: Environmental Refugees

1. J. T. Houghton, G. J. Jenkins, and J. J. Ephramus, eds., *Climate Change: The IPCC Scientific Assessment* (Cambridge, U.K.: Cambridge University Press, 1990).

2. D. Barker, *Environmental Migrants* (New York: United Nations Development Programme, 1989); J. L. Jacobson, *Environmental Refugees: A Yardstick of Habitability* (Washington, D.C.: Worldwatch Institute, 1989); D. Keen, *Refugees: Rationing the Right to Life—The Crisis in Emergency Relief* (London: Zed Books, 1992); A. Suhrke, *Environmental Degradation, Migration and Social Conflict* (Bergen, Norway: The Christian Nichelsen Institute, 1992); A. R. Zolberg, A. Suhrke, and S. Aguayo, *Escape from Violence: Conflict and the Refugee Crisis in the Developing World* (New York and London: Oxford University Press, 1989).

3. F. U. Mahtab, *Effect of Climate Change and Sea-Level Rise on Ban-*

gladesh (Dhaka, Bangladesh: Ministry of Agriculture, 1989); J. D. Milliman, J. M. Broadus, and F. Gable, "Environmental and Economic Implications of Rising Sea Level and Subsiding Deltas: The Nile and Bengal Examples," *Ambio*, 18(1989):340–345. For an updated and more precise analysis, see J. M. Broadus, "Possible Impacts of, and Adjustment to, Sea Level Rise: The Cases of Bangladesh and Egypt," in R. A. Warrick, E. M. Barrow, and T. M. L. Wigley, eds., *Climate and Sea Level Change: Observations, Projections, and Implications* (New York: Cambridge University Press, 1993), pp. 263–75; N. Myers, "Environmental Refugees: How Many Ahead?" *Bioscience* (in press, 1993).

4. K. A. Emanuel, "The Dependence of Hurricane Intensity on Climate," *Nature*, 326(1988):483–485; E. M. Rasmusson, "Potential Shifts of Monsoon Shifts Patterns Associated with Climate Warming," in J. C. Topping, ed., *Coping with Climate Change* (Washington, D.C.: The Climate Institute, 1989), 121–136.

5. F. U. Mahtab and A. Karim, "Land Use and Population Trends in Bangladesh," *Ambio*, 21(1, 1992):50–55; Milliman et al., "Nile and Bengal Examples."

6. Rasmusson, "Potential Shifts."

7. Houghton et al., *Climate Change* (1990); J. T. Houghton, B. A. Callender, and S. K. Varney, eds., *Climate Change 1992: The 1992 Supplementary Report to the IPCC Scientific Assessment* (New York: Cambridge University Press, 1992).

8. P. H. Gleick, "The Implications of Global Climatic Changes for International Security," *Climatic Change*, 15(1989):309–325; C. Rosenzweig and M. Parry, *Implications of Climate Change for International Agriculture: Global Food Trade and Vulnerable Regions* (Washington, D.C.: Environmental Protection Agency, 1992).

9. E. Badr and S. Darwish, *Estimation of Egyptian Population Beneath a Poverty Line in 1990/91* (Oxford: Environmental Change Unit, University of Oxford, 1991).

10. Broadus, "Possible Impacts of, and Adjustment to, Sea Level Rise: The Cases of Bangladesh and Egypt"; D. J. Stanley, "Subsidence in the Northeastern Nile Delta: Rapid Rates, Possible Causes and Consequences," *Science*, 240(1988):497–500.

11. M. El-Raey, "Responses to the Impacts of Greenhouse-Induced Sea-Level Rise on the Northern Coastal Regions of Egypt," in J. G. Titus, ed., *Changing Climate and the Coast* (Washington, D.C.: Environmental Protection Agency, 1990), Vol. 2, 225–238; M. El-Raey, S. Nasr, O. Frihy et al., *Quantitative Impact of Accelerated Sea-Level Rise Over Alexandria Governorate, Egypt*, cited in J. C. Topping, ed., *Implications of Climate Change and Sea Level Rise for Human Settlement—Examples from Rio de Janeiro, Shanghai, Hong Kong, Tokyo, Dakar and Alexandria* (Washington, D.C.:

The Climate Institute, 1992); O. E. Frihy, "Nile Delta Shoreline Change: Aerial Photographic Study of a 28-Year Period," *Journal of Coastal Research,* 4(1988):597–606; A. I. Kashef, "Salt-Water Intrusion in the Nile Delta," *Groundwater,* 21(1983):160–167; Milliman et al., "Nile and Bengal Examples."

12. Milliman et al., "Nile and Bengal Examples."

13. Delft Hydraulics, *Sea-Level Rise: A Worldwide Cost Estimate of Basic Coastal Defense Measures* (Delft, Netherlands: Delft Hydraulics, 1990); L. T. Edgerton, *The Rising Tide: Global Warming and World Sea Levels* (Washington, D.C.: Island Press, 1991); R. Frassetto, ed., *Impact of Sea Level Rise on Cities and Regions* (Venice: Marsilio Edetori, 1991); G. P. Heckstra, *Global Warming and Rising Sea Levels: The Policy Implications* (Delft, Netherlands: Delft Hydraulics Laboratory, 1990); J. L. Jacobson, "Holding Back the Sea," in L. R. Brown et al., eds., *State of the World 1990* (New York: W. W. Norton, 1990), 79–97; Topping, *Implications of Climate Change;* United Nations Environment Programme, *Criteria for Assessing Vulnerability to Sea-Level Rise: A Global Inventory for High-Risk Areas* (Nairobi: United Nations Environment Programme, 1989).

14. United Nations Environment Programme, *Assessing Vulnerability.*

15. Heckstra, *Global Warming.*

16. Edgerton, *Rising Tide;* Frassetto, *Sea Level Rise;* Heckstra, *Global Warming;* Jacobson, "Holding Back"; Topping, *Implications of Climate Change;* United Nations Environment Programme, *Assessing Vulnerability.*

17. R. Dolan and H. G. Goodell, "Sinking Cities," *American Scientist,* 74(1986):38–47; Topping, *Implications of Climate Change.*

18. Ibid.

19. United Nations Environment Programme, *Assessing Vulnerability.*

20. Ibid.

21. K. Voigt, *Global Ocean Observing System* (Paris: Intergovernment Oceanographic Commission, UNESCO, 1991).

22. J. G. Titus, ed., *Changing Climate and the Coast,* 2 vols. (Washington, D.C.: Environmental Protection Agency, 1990).

23. Heckstra, *Global Warming;* Titus, *Changing Climate;* United Nations Environment Programme, *Assessing Vulnerability.*

24. Voigt, *Global Ocean.*

25. Ibid.

26. Jacobson, "Holding Back"; R. J. Nicholls, K. C. Dennis, and S. P. Leatherman, *Impacts of Sea-Level Rise on Selected Developing Countries* (College Park, MD: Laboratory for Coastal Research, University of Maryland, 1990); Titus, *Changing Climate;* J. C. Topping, *Global Warming: Impact on Developing Countries* (Washington, D.C.: Overseas Development Council, 1990).

27. I. A. Sughandy, *Preliminary Findings on Potential Impact of Climate Change in Indonesia* (Jakarta, Indonesia: Ministry for Population and Environment, 1989).

28. M. Han, J. Hu, and L. Wu, *Adverse Impact of Projected One-Meter Sea Level Rise on China's Coastal Environment and Cities: A National Assessment* (College Park, MD: Center for Global Change, University of Maryland, 1990); B. Wang, C. Shenling, Z. Keqi et al., *Impacts of Sea Level Rise on the Shanghai Area*, cited in Topping, *Implications of Climate Change.*

29. Ibid.

30. V. Asthana, *National Assessment of Effects of a Possible Sea-Level Rise and Responses in India* (New Delhi: School of Environmental Sciences, Jawaharal Nehru University, 1989).

31. Topping, *Implications of Climate Change.*

32. Houghton et al., *Climate Change* (1990).

33. J. C. Topping, A. Qureshi, and S. A. Sherer, *Implications of Climate Change for the Asian Pacific Region* (Washington, D.C.: The Climate Institute, 1990).

34. G. Daily and P. R. Ehrlich, "An Exploratory Model of the Impact of Rapid Climate Change on the World Food Situation," *Proceedings of the Royal Society of London B,* 241(1990):232–244.

35. Ibid.

36. L. R. Brown et al., *State of the World 1992* (New York: W. W. Norton, 1992).

37. Rosenzweig and Parry, *Implications of Climate.*

38. M. Suliman, ed., *Greenhouse Effect and Its Impact on Africa* (London: Institute for African Alternatives, 1990).

39. S. F. Martin, "The Inhospitable Earth," *Refugees* (publication of the United Nations High Commission for Refugees), 89(1992):13–15.

CHAPTER 14: The Synergistic Connection

1. N. Myers, "Synergisms: Joint Effects of Climate Change and Other Forms of Habitat Destruction," in R. L. Peters and T. E. Lovejoy, eds., *Global Warming and Biological Diversity* (New Haven, CT: Yale University Press, 1992), 344–354.

2. National Research Council, *Ozone Depletion, Greenhouse Gases and Climate Change* (Washington, D.C.: National Academy Press, 1989); S. H. Schneider, *Global Warming: Are We Entering the Greenhouse Century?* (San Francisco: Sierra Club Books, 1989).

3. H. Rodhe and R. Herrera, *Acidification in Tropical Countries* (New York: John Wiley, 1988).

4. J. M. Dave, *Policy Options for Development in Response to Global At-*

mospheric Changes: Case Study for India for Greenhouse Effect Cases (New Delhi: Nehru University, 1988); M. Oppenheimer and R. H. Boyle, *Dead Heat: The Race against the Greenhouse Effect* (New York: Basic Books, 1990).

5. P. R. Ehrlich and A. H. Ehrlich, *The Population Explosion* (New York: Simon and Schuster, 1990).

6. A. B. Lovins and L. H. Lovins, "Least-Cost Climatic Stabilization," *Review of Energy and Environment*, 16(1991):433–531.

7. N. Myers, "Population/Environment Linkages: Discontinuities Ahead," *Ambio*, 21(1, 1992):116–118.

8. N. Myers, "Environmental Degradation and Some Economic Consequences in the Philippines," *Environmental Conservation*, 15(3, 1988):205–214.

9. N. Sadik, *Safeguarding the Future* (New York: United Nations Population Fund, 1990).

CHAPTER 15: Trade-offs with Military Security

1. R. L. Sivard, *World Military and Social Expenditures 1991* (Washington, D.C.: World Priorities, 1991).

2. Ibid.; United Nations Development Programme, *Human Development Report 1991* (New York: Oxford University Press, 1991).

3. Sivard, *Military and Social Expenditures.*

4. Sivard, *Military and Social Expenditures;* United Nations Development Programme, *Human Development Report 1992* (New York: Oxford University Press, 1992).

5. The Hunger Project, *How to Put an End to Hunger* (Providence, RI: The Hunger Project, Brown University, 1992).

6. M. S. Strong, Secretary-General's Speech to the U.N. Conference on Environment and Development, Rio de Janeiro, 4 June 1992.

7. U.S. Arms Control and Disarmament Agency, *World Military Expenditures and Arms Transfers* (Washington, D.C.: U.S. Government Printing Office, 1990).

8. R. S. McNamara, "Toward a New World Order," *EcoDecision,* 2(1991):14–19.

CHAPTER 16: The Policy Fallout

1. Brandt Commission, *North-South: A Program for Survival* (Report of the Independent Commission on International Development Issues) (New York: Pan Books, 1980).

2. J. MacNeill, P. Winsemius, and T. Yakushiji, *Beyond Interdependence: The Meshing of the World's Economy and the Earth's Ecology* (New York: Oxford University Press, 1991); S. Ramphal, *Our Country the Planet* (London: Lime Tree Press, 1992).

3. S. George, *The Debt Boomerang* (London: Pluto Press, 1992).

4. George, *Debt Boomerang*; H. O'Neill, ed., *Third World Debt: How Sustainable Are Current Strategies and Solutions?* (London: Cass Publishers, 1990).

5. P. Weaver, *Sustainable/Equitable Development and Debt* (Laxenburg, Austria: International Institute for Applied Systems Analysis, 1991).

6. N. Sadik, *Safeguarding the Future* (New York: United Nations Population Fund, 1990).

7. UNICEF, *State of the World's Children* (New York: UNICEF, 1990).

8. S. K. Tucker, *The Debt-Trade Linkage in U.S.–Latin American Trade* (Washington, D.C.: Overseas Development Council, 1990).

9. J. Wheeler, *Development Cooperation: Efforts and Policies of the Members of the Development Assistance Committee* (Paris: Organization for Economic Cooperation and Development, 1990).

10. P. Winglee, "Agricultural Trade Policies of Industrial Countries," *Finance and Development*, 26(1, 1989):9–12; J. Zietz and A. Valdes, *Costs of Protectionism to Developing Countries* (Washington, D.C.: The World Bank, 1986).

11. B. Coote, *The Trade Trap: Poverty and the Global Commodity Markets* (Oxford: Oxfam, 1992); Ramphal, *Our Country*; World Bank, *World Development Report 1992* (New York: Oxford University Press, 1992).

12. George, *Debt Boomerang*.

13. R. Repetto, "Deforestation in the Tropics," *Scientific American*, 262(1990):36–42.

14. George, *Debt Boomerang*.

15. P. M. Haas, *Saving the Mediterranean: The Politics of International Environmental Cooperation* (New York: Columbia University Press, 1990); M. Grenon and M. Batisse, *The Mediterranean Basin Blue Plan* (New York: Oxford University Press, 1991).

16. S. Keckes, "Achievements and Planned Development of UNEP's Regional Seas Programme and Comparable Programmes Sponsored by Other Parties," *UNEP Regional Seas Reports and Studies*, No. 1, 1982.

17. United Nations Environment Programme, *The State of the Marine Environment* (Nairobi: United Nations Environment Programme, 1990).

CHAPTER 17: Finale: A Personal Reflection

1. World Commission on Environment and Development, *Our Common Future* (Oxford: Oxford University Press, 1987).

2. N. Myers, "Tropical Deforestation and Species Extinctions: The Latest News," *Futures*, 17(1985):451–463.

Resource Guide

The reader who may be interested in following up on the ideas in this book can get in touch with any of the following organizations, all of which are concerned in one way or another with the overall theme of environmental security.

United States of America

Albert Schweitzer Institute, Wallingford, CT
Arms Control Association, Washington, D.C.
Bread for the World, Washington, D.C.
Carnegie Endowment for International Peace, Washington, D.C.
Carrying Capacity Network, Washington, D.C.
Center for Environmental Management and Fletcher School of Law and Diplomacy, Tufts University, Bedford, MA
Center for Global Change, University of Maryland, College Park, MD
Center for Defense Information, Washington, D.C.
Center for Immigration Policy and Refugee Assistance, Georgetown University, Washington, D.C.
Center for Peace Studies, University of Colorado, Boulder, CO
Center for Science and International Affairs, Harvard University, Cambridge, MA
Center for Strategic and International Studies, Georgetown University, Washington, D.C.

RESOURCE GUIDE

Center for War, Peace and the News Media, New York, NY
Coalition for New Foreign Policy, Washington, D.C.
Committee on International Security Studies, American Academy of Arts and Sciences, Cambridge, MA
Committee on Oceans, Environment and Science, National Security Council, The White House, Washington, D.C.
Council on Economic Priorities, New York, NY
Council on Foreign Relations, New York, NY
Environmental Security Network, University of Colorado, Boulder, CO
Foreign Policy Studies Program, Brookings Institution, Washington, D.C.
Global Education Associates, New York, NY
Global Security Program, The MacArthur Foundation, Chicago, IL
Indiana Center on Global Change and World Peace, Indiana University, Bloomington, IN
International Peace Research Association, Antioch College, Yellow Springs, OH
International Program of the Environment and Energy Studies Institute, Washington, D.C.
Institute for East-West Security Studies, New York, NY
Institute for Policy Studies, Washington, D.C.
Institute for Resource and Security Studies, Cambridge, MA
Institute on Global Conflict and Cooperation, University of California at San Diego, La Jolla, CA
National Commission for Economic Conversion and Disarmament, Washington, D.C.
National Peace Foundation, Washington, D.C.
Peace Through Law Education Fund, Washington, D.C.
Program on the Analysis and Resolution of Conflicts, Maxwell School of Citizenship and Public Affairs, Syracuse University, Syracuse, NY
Public Agenda Foundation, New York, NY
Pugwash USA, Washington, D.C.
Redefining National Security Program, Rocky Mountain Institute, Snowmass, CO
The Stanley Foundation, Muscatine, IA
U.S. Arms Control and Disarmament Agency, Washington, D.C.
Women's International League for Peace and Freedom, Washington, D.C.
Women's Action for Nuclear Disarmament Inc., Washington, D.C.
World Federalist Association, Washington, D.C.
World Policy Institute, New York, NY
World Priorities, Washington, D.C.

Canada

Canadian Institute for International Peace and Security, Ottawa, Ontario
Institute of Peace and Conflict Studies, Conrad Grebel College, Waterloo,
    Ontario
Peace and Conflict Studies Program, University of Toronto, Toronto

United Kingdom

Armament and Disarmament Unit, University of Sussex, Falmer, Sussex
Centre for International Peacebuilding, Chipping Norton, Oxon.
Centre for Security and Conflict Studies, Institute for the Study of Conflict,
    London
Global Commons Institute, London
Global Security Programme, University of Cambridge, Cambridge
International Centre for Peacebuilding, London
International Institute for Strategic Studies, London
Peace Research Institute, University of Bradford, Bradford
Pugwash, London
Richardson Institute for Peace Studies, University of Lancaster, Lancaster
Royal Institute for International Affairs, London
Saferworld Foundation, Bristol

Australia

Indian Ocean Centre for Peace Studies, University of Western Australia,
    Nedlands
Nautilus Pacific Research, North Fitzroy, Victoria

Germany

European Peace Research Association, Bonn
Frankfurt Foundation for Peace and Conflict Research, Frankfurt

India

Institute for Defense and Strategic Analysis, New Delhi

Japan

International Institute for Global Peace, Tokyo

Malaysia

Institute of Strategic and International Studies, Kuala Lumpur

Nigeria

African Peace Research Institute, Lagos

Norway

International Peace Research Institute, Oslo

Philippines

Peace Studies Institute, Manila

Former Soviet Union

Peace Research Institute, Russian Academy of Sciences, Moscow

Sweden

Life and Peace Institute, Uppsala
Stockholm International Peace Research Institute, Stockholm

Switzerland

Independent Commission for International Humanitarian Issues, Geneva

Index

INDEX

INDEX

Index

INDEX